# Foundations of Science Mathematics

# Foundations of Science Mathematics

SECOND EDITION

D. S. Sivia

J. L. Rhodes

S. G. Rawlings

OXFORD
UNIVERSITY PRESS

OXFORD
UNIVERSITY PRESS

Great Clarendon Street, Oxford, OX2 6DP,
United Kingdom

Oxford University Press is a department of the University of Oxford.
It furthers the University's objective of excellence in research, scholarship,
and education by publishing worldwide. Oxford is a registered trade mark of
Oxford University Press in the UK and in certain other countries

First Edition published in 1999
Second Edition published in 2020

Impression: 1

Published in the United States of America by Oxford University Press
198 Madison Avenue, New York, NY 10016, United States of America

British Library Cataloguing in Publication Data
Data available

Library of Congress Control Number: 2020938514

ISBN 978-0-19-879754-8

Printed in Great Britain by
Bell & Bain Ltd., Glasgow

# First edition prefaces

Mathematics plays a central role in the life of every scientist and engineer, from the earliest school days to late on into the college and professional years. While the subject may be met with enthusiasm or reluctance, all are required to learn it to some degree at university level. This Primer (**OCP 77**) begins by summarizing the basic concepts and results that should have been assimilated at high school, and goes on to extend the ideas to cover the material needed by the majority of undergraduates. While the emphasis may not be on the formal proof of each statement, the conveying of a meaningful understanding of where the results come from is at the heart of the book. It is hoped that the informality of the tutorial style adopted will endear the text to the novice, but that its concise nature will also ensure that it's a useful reference.

It is our pleasure to thank friends and colleagues who provided feedback on a draft of the early chapters: Drs. Richard Ibberson, Robert Leese, Jerry Mayers, Jeff Penfold, and David Waymont. Professor Richard Compton's guidance, encouragement, and patience was invaluable throughout. Finally, we gratefully acknowledge the deep influence that some mentors have had on our understanding and approach to mathematics; in particular, DSS would like to mention his father for the earliest exposure, Mr. Munson for his senior school days and Dr. Skilling while a maturing undergraduate.

*Oxford*                                                                                     D.S.S.
January, 1999                                                                       S.G.R.

This Primer (**OCP 82**) is the 'worked examples' half of a two-volume set that summarizes the basic mathematical concepts and results which should be familiar from high school, and then extends the ideas to cover the material needed by the majority of science undergraduates. It assumes a familiarity with the theoretical content of the subject explained in the main maths Primer (**OCP 77**), and aims to bolster the reader's practical confidence by providing 'model answers', and additional comments, for a wide range of exercises.

We would like to thank several friends and colleagues who showed interest in the project, and helped to facilitate its speedy conclusion: namely, Drs. Jerry Mayers, Jeff Penfold, and Andrew Taylor. Professor Richard Compton's guidance and encouragement was invaluable throughout.

*Oxford*                                                                                     D.S.S.
April, 1999                                                                           S.G.R.

# Preface to second edition

† www.oup.com/he/sivia-rhodes2e

The second edition sees the amalgamation of the original Primers into a single volume and the addition of a chapter on data analysis. Some changes have been made for improved clarity and the number of worked problems with a scientific context increased. The overall length of the text has been kept as short as possible, following the original philosophy, but more example questions are available at the online resource centre.[†]

Although the analysis of experimental data is central to the scientific method, it is often not taught well. This is partly due to the tortuous history of the subject, but mainly because a good understanding requires a knowledge of multivariate calculus and linear algebra. Since these are discussed prior to chapter 16, it enables us to introduce data analysis as a coherent and logical topic rather than a cookbook of statistical recipes.

We are grateful to Dr. David Waymont and and Prof. Richard Compton for useful comments on the new material, and to Adam Sills for his help in checking the accuracy of the equations. We are also indebted to Martha Bailes, Roseanna Levermore, Alice Roberts, Lydia Shinoj, Katherine Underhill and others for their professional assistance in the production of this book.

DSS would like to dedicate this edition to the memory of his longtime friend, and former coauthor, Steve Rawlings.

*Oxford*                                                                          D.S.S.
*Wakefield*                                                                      J.L.R.
September, 2020

# Contents

# Basic algebra and arithmetic

## 1.1 Elementary arithmetic

The first mathematical skill that we learn as children is arithmetic: starting with counting on our fingers, we soon progress to adding, taking away, multiplying, and dividing. All this seems quite straightforward until we encounter an expression like $2 + 6 \times 3$; does this mean that we add two to six and multiply the sum by three (to get 24), or should we multiply six by three and then add two (giving 20)? To avoid such ambiguities, an order ranking the priorities of the various basic arithmetical operations has been agreed; this convention can be remembered by the acronym BODMAS.

According to BODMAS, things occurring inside a bracket must be evaluated first; the next most important operation is 'orders', as in powers and roots; then it's division or multiplication, as they appear left-to-right, and lastly addition or subtraction. Thus our simple example above can be made explicit as follows

$$2 + 6 \times 3 = 2 + (6 \times 3) \neq (2 + 6) \times 3$$

*Brackets*
*Orders*
*Divide*
*Multiply*
*Add*
*Subtract*

While brackets are not necessary if the expression on the left is meant to stand for the one in the middle, they are essential for the one on the right.

Before moving on, we should also remind ourselves about the rule for multiplying out brackets; that is

$$(a + b)(c + d) = ac + ad + bc + bd \tag{1.1}$$

where $a$, $b$, $c$, and $d$ could be any arbitrary numbers (and a small space between quantities is equivalent to a multiplication, as in $ac = a \times c$).

## 1.2 Powers, roots, and logarithms

If a number, say $a$, is multiplied by itself, then we can write it as $a^2$ and call it $a$-squared. A triple product can be written as $a^3$ and is called $a$-cubed. In general, an N-times self-product is denoted by $a^N$ where the superscript, or *index*, N is referred to as the *power* of $a$.

Although the meaning of '$a$ to the power of N' is obvious when N is a positive integer (i.e. $1, 2, 3, \ldots$), what happens when it's zero or negative? This question is easily answered once we notice that the procedure for going from a power of

$a^1 = a$
$a^2 = a \times a$
$a^3 = a \times a \times a$
$a^4 = a \times a \times a \times a$

N to N−1 involves a division by $a$. Thus if $a^0$ is $a^1$ divided by $a$, then $a$ to the power of nought must be one (or *unity*); similarly, if $a^{-1}$ is $a^0$ divided by $a$, then it must be equal to one over $a$; and, in general, $a^{-N}$ is equivalent to the *reciprocal* of $a^N$. Thus, we have

$$a^0 = 1 \quad \text{and} \quad a^{-N} = \frac{1}{a^N} \tag{1.2}$$

The basic definition of powers leads immediately to the formula for adding the indices M and N when the numbers $a^M$ and $a^N$ are multiplied

$$a^M a^N = a^{M+N} \tag{1.3}$$

$$a^{1/2} a^{1/2} = a$$

$$a^{1/3} a^{1/3} a^{1/3} = a$$

While our discussion has so far focused only on the case of integer powers, suppose that we were to legislate that eqn (1.3) held for all values of M and N. Then, we would be led to the interpretation of fractional powers as *roots*. To see this, consider the case when M=N=1/2; it follows from eqn (1.3) that $a$ to the power of a half must be equal to the square root of $a$. Extending the argument slightly, if $a$ to the power of a third is multiplied by itself three times then we obtain $a$; therefore, $a^{1/3}$ must be equal to the cube root of $a$. In general, the $p^{th}$ root of $a$ is given by

$$a^{1/p} = \sqrt[p]{a} \tag{1.4}$$

where p is an integer. One final result on powers that we should mention is

$$\left(a^M\right)^N = a^{MN} \tag{1.5}$$

which can at least be verified readily for integer values of M and N. More complicated powers can be decomposed into a series of simpler manipulations by using the rules of eqns (1.2) – (1.5). For example

$$9^{-5/2} = \frac{1}{9^{5/2}} = \frac{1}{9^{2+1/2}} = \frac{1}{9^2 \, 9^{1/2}} = \frac{1}{81\sqrt{9}} = \frac{1}{243}$$

An alternative method of describing a number as a 'power of something' is to use *logarithms*. That is to say, if $y$ is written as $a$ to the power of $x$ then $x$ is the logarithm of $y$ to the *base* $a$

$$y = a^x \quad \Longleftrightarrow \quad x = \log_a(y) \tag{1.6}$$

$$\log_{10}(1) = \log(10^0) = 0$$
$$\log_{10}(10) = \log(10^1) = 1$$
$$\log_{10}(100) = \log(10^2) = 2$$

$$\ln(e^0) = 0$$
$$\ln(e^1) = 1$$
$$\ln(e^2) = 2$$

where the double-headed arrow indicates an equivalence, so that the expression on the left implies the one on the right and vice versa. Since we talk in powers of ten in everyday conversations (e.g. hundreds, thousands, millions), the use of $a=10$ is most common; this gives rise to the name *common* logarithm for $\log_{10}$, often abbreviated to just log (but this can be ambiguous). Other bases that are encountered frequently are 2 and 'e' (2.718 to 3 decimal places); we will meet the latter in more detail in later chapters, but state here that $\log_e$, or ln, is called the *natural* logarithm.

By combining the definition of the logarithm in eqn (1.6) with the rule of eqn (1.3), it can be shown that the 'log of a product is equal to the sum of the logs' (to any base); and, in conjunction with eqn (1.2), that the 'log of a quotient is equal to the difference of the logs'

$$\log(AB) = \log(A) + \log(B) \quad \text{and} \quad \log(A/B) = \log(A) - \log(B) \quad (1.7)$$

Similarly, eqn (1.6) allows us to rewrite eqn (1.5) in terms of the log of a power and to derive a formula for changing the base of a log (from $a$ to $b$)

$$\log(A^{\beta}) = \beta \log(A) \quad \text{and} \quad \log_b(A) = \log_a(A) \times \log_b(a) \quad (1.8)$$

Thus $\ln(x) = 2.303 \log(x)$, for example, where the numerical prefactor is $\ln(10)$ to four significant figures.

## 1.3  Quadratic equations

The simplest type of equation involving an 'unknown' variable, say $x$, is one that takes the form $ax + b = 0$, where $a$ and $b$ are (known) constants. Such *linear* equations can easily be rearranged according to the rules of elementary algebra, 'whatever you do to one side of the equation, you must do exactly the same to the other', to yield the solution $x = -b/a$.

A slightly more complicated situation, which is met frequently, is that of a *quadratic* equation; this takes the general form

$$ax^2 + bx + c = 0 \quad (1.9)$$

The crucial difference between this and the linear case is the occurrence of the $x^2$ term, which makes it far less straightforward to work out which values of $x$ satisfy the equation. If we could rewrite eqn (1.9), albeit divided by $a$, as

$$(x - x_1)(x - x_2) = 0$$

where $x_1$ and $x_2$ are constants, then the solutions are obvious: either $x = x_1$ or $x = x_2$, because the product of two numbers can only be zero if either one or the other (or both) is nought. While such a *factorization* may not be easy to spot, it is not difficult to rearrange eqn (1.9) into the form

$$(x + \alpha)^2 - \beta = 0$$

a procedure called 'completing the square', where $\alpha$ and $\beta$ can be expressed in terms of the constants $a$, $b$, and $c$ (but don't involve $x$). This leads to the following general formula for the two solutions of a quadratic equation

$$x = \frac{-b \pm \sqrt{b^2 - 4ac}}{2a} \quad (1.10)$$

$$x_1 x_2 = c/a$$

$$x_1 + x_2 = -b/a$$

$$\alpha = \frac{b}{2a}, \quad \beta = \frac{b^2}{4a^2} - \frac{c}{a}$$

and is equivalent to $x = -\alpha + \sqrt{\beta}$ and $x = -\alpha - \sqrt{\beta}$. Since the square of any number, positive or negative, is always greater than or equal to zero, we require that $b^2 \geqslant 4ac$ for eqn (1.10) to yield 'real' values of $x$.

## 1.4 Simultaneous equations

There are many situations in which an equation will contain more than one variable; taking the case of just two, say $x$ and $y$, a simple example would be $x + y = 3$. On its own, this relationship does not determine the values of $x$ and $y$ uniquely. Indeed, there are an infinite number of solutions which satisfy the equation: $x = 0$ and $y = 3$, or $x = 1$ and $y = 2$ or, more generally, $x = p$ and $y = 3 - p$. To pin down $x$ and $y$ to a single possibility, we need one more equation to constrain them; e.g. $x - y = 1$. The two conditions can only be satisfied at the same time if $x = 2$ and $y = 1$, and are an example of solving *simultaneous* equations.

The easiest simultaneous equations are linear ones, where the variables only appear as separate entities to their first power (perhaps multiplied by a constant) all added together. The simplest of these is a two-by-two system, whose general from can be written as

$$a x + b y = \alpha$$
$$c x + d y = \beta$$

$$x = \frac{\alpha d - \beta b}{a d - b c}$$

$$y = \frac{\beta a - \alpha c}{a d - b c}$$

This can be solved by substituting for $x$, or $y$, from one of the equations into the other. For example, the first equation gives $y = (\alpha - a x)/b$; putting this in for $y$ into the second equation yields a linear relationship for $x$; thus we readily obtain $x$, and hence $y$. An extension of this procedure allows us to determine the values of three variables ($x$, $y$, and $z$) uniquely when given three linear simultaneous equations, and so on. The only proviso is that all the equations must be genuinely distinct; that is, we cannot repeat the same one twice, or generate a third by combining any two (or more).

We should also note that a unique solution is only guaranteed if all the simultaneous equations are linear. If the second one in our example at the start of this section had been $x^2 - y = 3$, then the substitution of $y$ from $x + y = 3$ would give the quadratic equation $x^2 + x - 6 = 0$. Factorization of this as $(x + 3)(x - 2) = 0$ tells us that either $x = 2$ or $x = -3$, and hence that $y = 1$ or $y = 6$ respectively.

## 1.5 The binomial expansion

It is easily shown, by putting $c = a$ and $d = b$ in eqn (1.1), that the square of the sum of two numbers can be written as

$$(a+b)^2 = a^2 + 2ab + b^2$$

Multiplying this by the sum again yields the following for the cube

$$(a+b)^3 = a^3 + 3a^2b + 3ab^2 + b^3$$

When this procedure is repeated many times, a systematic pattern emerges for the $N^{\text{th}}$ power of $a+b$; it is called the *binomial* expansion

$$(a+b)^N = \sum_{r=0}^{N} {}^NC_r\, a^r\, b^{N-r} \qquad (1.11)$$

The capital Greek symbol $\Sigma$ stands for a 'sum from $r=0$ to $r=N$', through the integers $r=1, 2, 3, \cdots, N-1$. The binomial coefficients, ${}^NC_r$, which are often denoted by an alternative long bracket notation, are defined by

$$\sum_{r=0}^{N} A_r = A_0 + A_1 + \cdots + A_N$$

$${}^NC_r = \binom{N}{r} = \frac{N!}{r!\,(N-r)!} \qquad (1.12)$$

where the *factorial* function is given by the product

$$N! = N \times (N-1) \times (N-2) \times \cdots \times 3 \times 2 \times 1 \qquad (1.13)$$

Although not obvious from eqn (1.13), we will see later that $0!=1$.

Rather than using eqn (1.12), the coefficients in a binomial expansion can also be ascertained with a *Pascal's triangle*. In this, apart from the ones down the edges, each number is generated by adding the two closest neighbours in the line above. The 1 2 1 row of the triangle then gives the coefficients for $N=2$, the 1 3 3 1 row corresponds to $N=3$, and so on.

```
              1
            1   1
          1   2   1
        1   3   3   1
      1   4   6   4   1
    1   5  10  10   5   1
```

## 1.6 Arithmetic and geometric progressions

Sometimes we need to work out the sum of a sequence of numbers that are generated according to a certain algebraic rule. The simplest example is that of an *arithmetic progression*, or AP, where each term is given by the previous one plus a constant

$$a + (a+d) + (a+2d) + (a+3d) + \cdots + (l-d) + l$$

If there are N terms, then the last one, $l$, is related to the first, $a$, through the *common difference*, $d$, by $l = a + (N-1)d$. The addition of the above series with a copy of itself written in reverse order shows that twice the sum which we seek is equal to N times $a + l$; hence, the formula for the sum of an AP is

$$1 + 2 + 3 + \cdots + N = \frac{N(N+1)}{2}$$

$$\sum_{j=1}^{N} a + (j-1)d = \frac{N}{2}\left[2a + (N-1)d\right] \qquad (1.14)$$

Another case which is met frequently is that of a *geometric progression*, or GP, where each term is given by the previous one times a constant

$$a + ar + ar^2 + ar^3 + \cdots + ar^{N-2} + ar^{N-1}$$

If we subtract from the above series a copy of itself that has been multiplied by the *common ratio* $r$, then it can be shown that $1-r$ times the sum that we seek is equal to $a - ar^N$; hence, the formula for the sum of a GP is

$$\sum_{j=1}^{N} ar^{j-1} = \frac{a\left(1-r^N\right)}{1-r} \qquad (1.15)$$

$$1 - \frac{1}{2} + \frac{1}{4} - \frac{1}{8} + \cdots = \frac{2}{3}$$

If the *modulus* of the common ratio is less than unity, so that $-1 < r < 1$, then the sum of the GP does not 'blow up' as the number of terms becomes infinite. Indeed, since $r^N$ becomes negligibly small as N tends to infinity, eqn (1.15) simplifies to

$$\sum_{j=1}^{\infty} a\, r^{j-1} = \frac{a}{1-r} \qquad \text{for } |r| < 1 \qquad (1.16)$$

## 1.7 Partial fractions

For the final topic in this chapter, we turn to the subject of *partial fractions*. These are best explained by the use of specific examples, such as

$$\frac{5x+7}{(x-1)(x+3)} = \frac{3}{(x-1)} + \frac{2}{(x+3)}$$

$$\frac{1}{2} - \frac{1}{3} = \frac{3}{6} - \frac{2}{6} = \frac{1}{6}$$

The act of combining two, or more, fractions into a single entity is familiar to us, through 'finding the lowest common denominator', and is usually referred to as 'simplifying the equation'. Here we are interested in the reverse procedure of decomposing a fraction into the sum of several constituent parts. Although this may sound like a backwards step, it sometimes turns out to be a very useful manipulation.

The situations in which partial fractions arise involve the ratios of two *polynomials* (the sums of positive integer powers of a variable, like $x$), where the denominator can be written as the product of simpler components. We must first check that the *degree* (or highest power of $x$) of the numerator is less than that of the denominator, and 'divide out' if it is not; this is analogous to rewriting a top-heavy fraction as a whole number plus a proper fraction (e.g. $7/3 = 2 + 1/3$). If the numerator in our earlier example had been $2x^2 + 9x + 1$, then we would divide it by the denominator $x^2 + 2x - 3$ (when multiplied out), by 'long division' if necessary, to obtain

$$x^2 + 2x - 3 \overline{\smash{\big)}\ 2x^2 + 9x + 1}$$
$$\underline{2x^2 + 4x - 6}$$
$$5x + 7$$

$$\frac{2x^2 + 9x + 1}{(x-1)(x+3)} = 2 + \frac{5x+7}{(x-1)(x+3)}$$

The second part on the right can then be split up into partial fractions, as indicated at the start of this section; let's consider the mechanistic details of how this is done.

Our example constitutes the simplest case, where the denominator is a product of linear terms. This can always be expressed as

$$\frac{5x+7}{(x-1)(x+3)} = \frac{A}{(x-1)} + \frac{B}{(x+3)}$$

where A and B are constants. If there had been a third term on the bottom, $2x + 3$ say, there would be an additional fraction on the right, $C/(2x+3)$. To evaluate A and B (and C etc.), we can rearrange the right-hand side to have the same common denominator as on the left, and then equate the numerators

$$5x + 7 = A(x+3) + B(x-1)$$

For this to be satisfied, the coefficients of the various powers of $x$ on both sides must be identical; this leads to a set of simultaneous equations for the desired constants. Alternatively, specific values of $x$ can be substituted into the equation. Putting $x = 1$ gives $4A = 12$, for example, and setting $x = -3$ gives $4B = 8$. This approach leads to a procedure known as the 'cover-up' rule: to evaluate $A$, cover up $x - 1$ in the original expression and substitute $x - 1 = 0$ in what remains; for $B$, cover up $x + 3$ and put $x + 3 = 0$ in the rest.

$$A + B = 5$$
$$3A - B = 7$$

If there had been a quadratic factor in the denominator of our example, then its partial component would require a linear term in its numerator

$$\frac{5x + 7}{(x - 1)(x^2 + 4x - 3)} = \frac{A}{(x - 1)} + \frac{Bx + C}{(x^2 + 4x - 3)}$$

Apart from this generalization, that an $N^{th}$ degree factor on the bottom needs an $(N-1)^{th}$ degree polynomial on the top, the procedure for calculating the associated constants is the same: rearrange the right-hand side to have the same common denominator as the left and equate numerators.

$$5x + 7 = A(x^2 + 4x - 3) + (Bx + C)(x - 1)$$

$A$ is evaluated most easily by putting $x = 1$, giving $8A = 12$, and is equivalent to using the cover-up rule. $B$ and $C$ can then be ascertained by equating the coefficients of the $x^2$ and $x^0$ terms on both sides of the equation (and confirmed by those of $x^1$).

$$A + B = 0$$
$$4A - B + C = 5$$
$$3A - C = 7$$

The final case of note concerns the occurrence of repeated factors in the denominator, such as $(x + 3)^2$. We could expand this as $x^2 + 6x + 9$ and use the procedure for dealing with a quadratic discussed above, but a more useful decomposition tends to be

$$\frac{5x + 7}{(x - 1)(x + 3)^2} = \frac{A}{(x - 1)} + \frac{B}{(x + 3)} + \frac{C}{(x + 3)^2}$$

The related constants can be evaluated in the usual manner, by equating the numerators after the denominators have been made equal

$$5x + 7 = (x + 3)\left[A(x + 3) + B(x - 1)\right] + C(x - 1)$$

$A$ and $C$ are given readily by putting $x = 1$ and $x = -3$ or, equivalently, by the cover-up rule; $B$ then follows for the coefficients of $x^2$, or $x^1$, or $x^0$.

## 1.8 Worked examples

**(1)** Evaluate (a) $4^{3/2}$, (b) $27^{-2/3}$, (c) $\log_2(8)$ and (d) $\log_2(8^3)$.

(a) $4^{3/2} = 4^{1/2 \times 3} = \left(4^{1/2}\right)^3 = \left(\sqrt{4}\right)^3 = 2^3 = \underline{8}$

$or = 4^{1 + 1/2} = 4^1 \, 4^{1/2} = 4\sqrt{4} = 4 \times 2 = \underline{8}$

(b) $27^{-2/3} = \dfrac{1}{27^{2/3}} = \dfrac{1}{\left(\sqrt[3]{27}\right)^2} = \dfrac{1}{3^2} = \underline{\dfrac{1}{9}}$

(c) $\log_2(8) = \log_2(2^3) = \underline{3}$

(d) $\log_2(8^3) = 3\log_2(8) = 3\times3 = \underline{9}$

or $= \log_2\left[(2^3)^3\right] = \log_2(2^9) = \underline{9}$

---

**(2)** The pH scale of acidity is a logarithmic measure of the (molar) concentration of hydrogen ions present in an aqueous solution: $\mathrm{pH} = -\log_{10}[H^+]$. Calculate $[H^+]$, in moles per litre, for solutions with pH values of (a) 4.82, (b) 10.47, (c) 0.81 and (d) $-0.66$.

---

$$\mathrm{pH} = -\log_{10}\left[H^+\right] \iff \left[H^+\right] = 10^{-\mathrm{pH}}$$

Therefore the hydrogen ion concentrations, in $\underline{\mathrm{mol\ dm^{-3}}}$, are

(a) $10^{-4.82} = \underline{1.51\times10^{-5}}$

(b) $10^{-10.47} = \underline{3.39\times10^{-11}}$

(c) $10^{-0.81} = \underline{0.155}$

(d) $10^{0.66} = \underline{4.57}$

---

**(3)** Solve (a) $x^2-5x+6=0$, (b) $3x^2+5x-2=0$, (c) $x^2-4x+2=0$.

---

(a) $x^2-5x+6 = (x-3)(x-2) = 0 \qquad \therefore \underline{x=2 \text{ or } x=3}$

(b) $3x^2+5x-2 = (3x-1)(x+2) = 0 \quad \therefore \underline{x=-2 \text{ or } x=1/3}$

If this factorization is difficult to spot, then the result could be obtained from the general formula of eqn (1.10) with $a=3$, $b=5$, and $c=-2$.

$$x = \frac{-5\pm\sqrt{25+24}}{6} = \frac{-5\pm7}{6} = -\frac{12}{6} \text{ or } \frac{2}{6}$$

(c) $x = \dfrac{4\pm\sqrt{16-8}}{2} = 2\pm\dfrac{\sqrt8}{2} = 2\pm\sqrt{\dfrac{8}{4}} \qquad \text{i.e. } \underline{x=2\pm\sqrt2}$

---

**(4)** The degree of dissociation of an acid, $\alpha$ (between 0 and 1), is related to its molar concentration, $c_0$, and dissociation constant, $K_a$, through Ostwald's dilution law: $(1-\alpha)K_a = \alpha^2 c_0$. Calculate $\alpha$ for a medium strength acid of concentration 0.01 mol dm$^{-3}$ and $K_a = 0.1$ mol dm$^{-3}$. With $[H^+] = \alpha c_0$, what is the pH?

$$c_0 \alpha^2 + K_a \alpha - K_a = 0$$

$$\therefore \quad \alpha = \frac{-K_a \pm \sqrt{K_a^2 + 4K_a c_0}}{2c_0} = \frac{-0.1 \pm \sqrt{0.01 + 0.004}}{0.02} = -5 \pm 5.92$$

$$\therefore \quad \underline{\alpha = 0.92} \quad \text{because } 0 \leqslant \alpha \leqslant 1$$

$$pH = -\log_{10}[H^+] = -\log_{10}(0.0092) = \underline{2.04}$$

---

**(5)** Solve the following simultaneous equations:

(a) $\begin{aligned} 3x + 2y &= 4 \\ x - 7y &= 9 \end{aligned}$    (b) $\begin{aligned} x^2 + y^2 &= 2 \\ x - 2y &= 1 \end{aligned}$    (c) $\begin{aligned} 3x + 2y + 5z &= 0 \\ x + 4y - 2z &= 9 \\ 4x - 6y + 3z &= 3 \end{aligned}$

---

(a)  $3x + 2y = 4 \quad - (1)$

$\quad x - 7y = 9 \quad - (2)$

$\qquad (1) - 3 \times (2) \quad \Rightarrow \quad 3x + 2y - 3(x - 7y) = 4 - 27$

$\qquad\qquad \therefore \quad 23y = -23$

$\qquad\qquad \therefore \quad y = -1, \quad \text{and} \quad (2) \Rightarrow x + 7 = 9$

$\qquad\qquad \text{i.e.} \quad \underline{x = 2 \quad \text{and} \quad y = -1}$

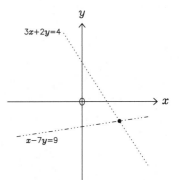

(b)  $x^2 + y^2 = 2 \quad - (3)$

$\quad x - 2y = 1 \quad - (4)$

Substituting $x = 2y + 1$ from (4) into (3):

$\qquad (2y + 1)^2 + y^2 = 2$

$\therefore \quad 4y^2 + 4y + 1 + y^2 = 2$

$\qquad \therefore \quad 5y^2 + 4y - 1 = 0$

$\qquad \therefore \quad (5y - 1)(y + 1) = 0 \quad \Rightarrow \quad y = \frac{1}{5} \text{ or } y = -1$

When $y = 1/5$,   (4) $\Rightarrow x = 1 + 2/5$

When $y = -1$,   (4) $\Rightarrow x = 1 - 2$

Hence  $\underline{x = 7/5 \text{ and } y = 1/5, \text{ or } x = -1 \text{ and } y = -1}$

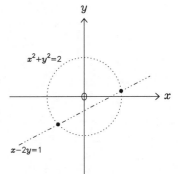

(c)  $3x + 2y + 5z = 0 \quad - (5)$

$\quad x + 4y - 2z = 9 \quad - (6)$

$\quad 4x - 6y + 3z = 3 \quad - (7)$

$$(5) - 3 \times (6) \quad \Rightarrow \quad -10y + 11z = -27$$
$$(7) - 4 \times (6) \quad \Rightarrow \quad -22y + 11z = -33$$
$$\left. \right\} \quad \begin{array}{l} y = 1/2 \\ z = -2 \end{array}$$

$$\therefore \quad \underline{x = 3 \,, \quad y = 1/2 \quad \text{and} \quad z = -2}$$

---

**(6)** The relative atomic mass of titanium, $A_r$, is 47.9183. If the isotopic abundances are

| Isotopic mass | 46 | 47 | 48 | 49 | 50 |
|---|---|---|---|---|---|
| % abundance | 8.25 | $x$ | 73.72 | 5.41 | $y$ |

calculate $x$ and $y$, the percentage abundance of $^{47}$Ti and $^{50}$Ti.

---

$$8.25 + x + 73.72 + 5.41 + y = 100$$

$$\therefore \quad x + y = 12.62 \quad - (1)$$

$$46 \times 8.25 + 47x + 48 \times 73.72 + 49 \times 5.41 + 50y = 100\,A_r$$

$$\therefore \quad 379.50 + 47x + 3538.56 + 265.09 + 50y = 4791.83$$

$$\therefore \quad 47x + 50y = 608.68 \quad - (2)$$

Substituting $y = 12.62 - x$ from (1) into (2):

$$47x + 631.00 - 50x = 608.68$$

$$\therefore \quad -3x = -22.32$$

$$\therefore \quad \underline{x = 7.44} \,, \quad \text{and} \quad (1) \Rightarrow \underline{y = 5.18}$$

---

**(7)** Decompose the following into partial fractions:

(a) $\dfrac{1}{x^2 - 5x + 6}$    (b) $\dfrac{x^2 - 5x + 1}{(x-1)^2 (2x-3)}$    (c) $\dfrac{11x + 1}{(x-1)(x^2 - 3x - 2)}$

---

(a) $\quad \dfrac{1}{x^2 - 5x + 6} = \dfrac{1}{(x-3)(x-2)} = \dfrac{A}{(x-3)} + \dfrac{B}{(x-2)}$

$$\therefore \quad 1 = A(x-2) + B(x-3)$$

Putting $x = 2 \Rightarrow B = -1$,    and    $x = 3 \Rightarrow A = 1$

i.e. $\quad \dfrac{1}{x^2 - 5x + 6} = \dfrac{1}{(x-3)} - \dfrac{1}{(x-2)}$

We could, in fact, have written this answer straight down by using the 'cover-up' rule.

(b) $\dfrac{x^2-5x+1}{(x-1)^2(2x-3)} = \dfrac{A}{(x-1)^2} + \dfrac{B}{(x-1)} + \dfrac{C}{(2x-3)}$

$\therefore\ x^2-5x+1 = A(2x-3) + (x-1)\big[B(2x-3) + C(x-1)\big]$

Putting $x=1 \Rightarrow A = 3$, and $x=3/2 \Rightarrow C = -17$

Equating coefficients of $x^2 \Rightarrow 1 = 2B + C \quad \therefore B = 9$

i.e. $\dfrac{x^2-5x+1}{(x-1)^2(2x-3)} = \dfrac{3}{(x-1)^2} + \dfrac{9}{(x-1)} - \dfrac{17}{(2x-3)}$

In this question, A and C are readily obtained with the 'cover-up' rule; B, however, requires the formal analysis.

(c) $\dfrac{11x+1}{(x-1)(x^2-3x-2)} = \dfrac{A}{(x-1)} + \dfrac{Bx+C}{(x^2-3x-2)}$

$\therefore\ 11x+1 = A(x^2-3x-2) + (x-1)(Bx+C)$

Putting $x=1 \Rightarrow A = -3$

Putting $x=0 \Rightarrow 1 = -2A - C \quad \therefore C = 5$

Equating coefficients of $x^2 \Rightarrow 0 = A + B \quad \therefore B = 3$

i.e. $\dfrac{11x+1}{(x-1)(x^2-3x-2)} = \dfrac{3x+5}{(x^2-3x-2)} - \dfrac{3}{(x-1)}$

Here only A follows from the 'cover-up' rule; both B and C need the more formal analysis.

---

**(8)** Evaluate the infinite summation $\displaystyle\sum_{n=0}^{\infty} e^{-\beta(n+1/2)}$ where $\beta > 0$.

---

$\displaystyle\sum_{n=0}^{\infty} e^{-\beta(n+1/2)} = e^{-\beta/2} + e^{-3\beta/2} + e^{-5\beta/2} + e^{-7\beta/2} + \cdots$

$= $ sum of infinite GP with $a = e^{-\beta/2}$ and $r = e^{-\beta}$

$= \dfrac{e^{-\beta/2}}{1 - e^{-\beta}}$

While the above example was merely an exercise in recognizing and evaluating the sum of an infinite GP, it does relate to a problem of physical interest. In quantum mechanics, the solutions of the Schrödinger equation for a particle in a harmonic potential (such as that found for a diatomic molecule) show that the energy levels for the system are given by

$$E_n = \left(n + \tfrac{1}{2}\right) h\nu$$

where $n = 0, 1, 2, 3, \ldots$, $h$ is *Planck's constant* and $\nu$ is the natural frequency of the vibrations given by the curvature of the potential well; the case of $n = 0$ is known as the *ground state*, with a *zero-point* energy of $E_0 = h\nu/2$. The probability that a particle occupies an energy level $E_n$ is given by the *Boltzmann factor* $\exp(-E_n/kT)$, where $k$ is the *Boltzmann constant* and $T$ is the temperature (in Kelvin). The sum of these occupation probabilities yields the summation above, with $\beta = h\nu/kT$, and is known as the *partition function*.

# 2  Curves and graphs

## 2.1  Straight lines

The relationship between two quantities, generically called $x$ and $y$, is often best understood when displayed in the form of a *graph*. That is when linked pairs of $x$ and $y$ values are plotted as distinct points on a piece of paper, and joined together to form a curve. The horizontal and vertical displacements, relative to a reference location for $x=0$ and $y=0$ known as the *origin*, give the *coordinates* $(x, y)$ of the points; by convention, $x$ increases from left to right and $y$ from bottom to top. With such a visual manifestation, it is easy to see (literally) how $y$ changes with $x$. The simplest variation is none at all! This corresponds to the equation $y=c$, and yields a graph consisting of just a horizontal line whose vertical position is determined by the constant $c$.

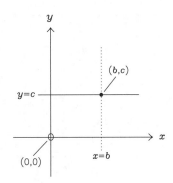

A more general case of a straight line is one that is tilted. This takes the algebraic form

$$y = m\,x + c \qquad\qquad (2.1)$$

where $m$ is a constant that controls the magnitude and sense of the slope. If $m$ is positive, then $y$ increases with $x$; if it's negative, then $y$ decreases with $x$; and if $m=0$, then we return to the situation of no variation. The values of $m$ and $c$ can be obtained from any two points, $(x_1, y_1)$ and $(x_2, y_2)$, on the line

$$m = \frac{y_2 - y_1}{x_2 - x_1} \quad \text{and} \quad c = \frac{x_2\,y_1 - x_1\,y_2}{x_2 - x_1}$$

and are usually referred to as the *gradient* and *intercept*, respectively. The latter is also equal to the value of $y$ when $x=0$, so that changing $c$ moves the line up and down; the corresponding point on the $x$-axis, when $y=0$, is called the *abscissa*.

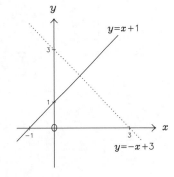

In the light of this discussion, we can interpret the (unique) solution of the $2 \times 2$ set of linear simultaneous equations in section 1.4 as being given by the intersection of two straight lines.

## 2.2  Parabolas

In chapter 1, we came across quadratic equations; these involved an $x^2$ term, in addition to the linear ones. Their general form can be written as

$$y = a\,x^2 + b\,x + c \tag{2.2}$$

and trace out a *parabola*, like the trajectory of a cannon ball, when plotted as a graph. The sign of the $x^2$ coefficient, $a$, determines whether the curve bends upwards or downwards at the ends, and its magnitude controls how quickly it does so. To understand how $a$, $b$, and $c$ affect the location of the parabola, it is best to rewrite eqn (2.2) by 'completing the square' as in section 1.3

$$y = a\,(x + \alpha)^2 + \gamma$$

It is then easy to see that the *turning point* of the curve (the apex where it bends over) occurs at $x = -\alpha$, or a half of minus $b/a$, and $y = \gamma$, which is $c$ minus a quarter of $b^2/a$.

When solving a quadratic equation, we seek the values of $x$ which make eqn (2.2) equal to zero. Graphically, these corresponds to the two points where the parabola cuts the $x$-axis; they become coincident when $b^2 = 4\,a\,c$, however, because the turning point then occurs at $y = 0$. There are situations when the curve never crosses the $x$-axis, of course, so that no (real) solutions exist; this is the case of section 1.3 when $b^2 < 4ac$.

The nature of eqn (2.2) also explains why we do not obtain a unique answer for a $2 \times 2$ set of simultaneous equations when one of them is quadratic and the other linear: we are looking for the intersections between a parabola and a straight line.

## 2.3 Polynomials

Both the straight line and the parabola are special cases of a family of curves known as *polynomials*. The similarity between them is more obvious from the algebraic form of the equation that defines them

$$y = a_0 + a_1\,x + a_2\,x^2 + a_3\,x^3 + \cdots + a_N\,x^N \tag{2.3}$$

where $a_0, a_1, a_2, \cdots, a_N$, are constants; the highest power of $x$, N, is known as the *degree*, or *order*, of the polynomial. If all the coefficients except $a_0$ are zero, or put simply N = 0, we obtain a horizontal line; with N = 1, we recover the two components needed for a general straight line; and a parabola emerges if N = 2.

After the linear case, the graphs of the polynomials start to develop bends and wiggles in them. At the ends, where the magnitude of $x$ is very large, all the terms in eqn (2.3) become negligible in comparison to the last one. In the extremities, therefore, the polynomial curves upwards or downwards with the same sense if the degree N is even, and in opposing ways if N is odd; while the amount of the curvature increases with N, its absolute rate and direction is governed by the size and sign of the coefficient $a_N$ respectively. On the journey from its starting point (when $x \to -\infty$) to its final destination ($x \to \infty$), the $y$ coordinate can go through a number of oscillations at intermediate values of $x$: an N$^{th}$ order polynomial can have up to N $-1$ turning points, but it may have fewer by multiple factors of two.

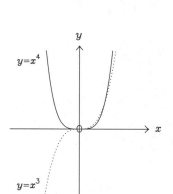

$y$

$y = x^2 - 2x + 2$

$y = -x^2 + 2x + 2$

$x$

$y$

$y = x^4$

$y = x^3$

$x$

$y$

$y = x^4 + x^3 - 2x^2 + 1$

$y = x^3 - x$

$x$

A *cubic* (N$=$3) can either have two turning points (a 'maximum' and a 'minimum') or none at all, for example, and a *quartic* (N$=$4) may have three or just one.

The values of $x$ which make eqn (2.3) equal to zero are called the *roots* of the polynomial. Graphically, they correspond to the points where the curve crosses the $x$-axis. Given the discussion above, a little thought shows that a polynomial of degree N can have up to N 'real' roots.

## 2.4 Powers, roots, and logarithms

One of the most basic relationships between two quantities is that of a *proportionality*. If $y$ becomes twice as large when $x$ doubles, and three times as big when $x$ trebles, and so on, then we say that '$y$ is directly proportional to $x$' and write it as $y \propto x$. This can be turned into an equation by introducing a multiplicative constant $k$, so that $y = kx$, and is represented graphically by a straight line that passes through the origin. Generalizing from the linear case, $y$ may be proportional to any power of $x$

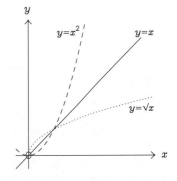

$$y \propto x^{\beta} \iff y = k x^{\beta} \qquad (2.4)$$

If $\beta = 2$, for example, then $y$ increases four-fold when $x$ doubles and nine-fold when $x$ trebles; by contrast, $y$ is only twice as large when $x$ is four-times bigger, and merely trebles when $x$ increases nine-fold, if $\beta = 1/2$.

Although the algebraic form of eqn (2.4) is very simple, it is not easy to ascertain the values of $\beta$ and $k$ from a graph of $y$ against $x$; this is because, other than when $\beta$ is 0 or $\pm 1$, it is difficult to judge by eye a curve that has a varying degree of bend. We can make the analysis much more straightforward, however, by taking the logarithm of eqn (2.4)

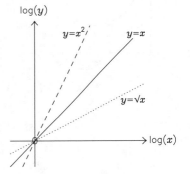

$$\log(y) = \beta \log(x) + \log(k) \qquad (2.5)$$

where we have used the results from eqns (1.7) and (1.8) to expand $\log(k\,x^{\beta})$ on the right-hand side. Thus if we plot $\log(y)$ against $\log(x)$, to any base, for $x$ (and $k$) greater than zero, we will obtain a straight line whose gradient and intercept are given by $\beta$ and $\log(k)$ respectively.

Another power-related equation which crops up frequently in theoretical work is that of an *exponential decay*

$$y = A\,e^{-\lambda x} = A\,\exp(-\lambda x) \qquad (2.6)$$

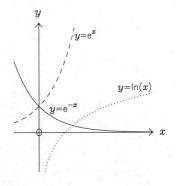

where 'exp' is an alternative notation for 'e to the power of', and A and $\lambda$ are constants. Strictly speaking, $\lambda$ must be positive for a decay; otherwise, there is exponential growth. As in the earlier situation, eqn (2.6) can be turned into a simpler graphical form by taking its logarithm (preferably to base e)

$$\ln(y) = \ln(A) - \lambda\,x \qquad (2.7)$$

where we have used the definition of $\log_e$ from eqn (1.6) in writing the last term. In other words, if we plot $\ln(y)$ against $x$ (for $A > 0$) then we obtain a straight line with a gradient and intercept given by $-\lambda$ and $\ln(A)$ respectively.

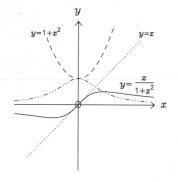

The graphs of complicated functions of polynomials, and exponentials, can often be sketched by first considering the behaviours of their constituent parts. For example, $y = x/(1+x^2)$ can be thought of in the following way: (i) $1+x^2$ is a parabola that is large and positive when $x \to \pm\infty$, and has a minimum at $x = 0$ and $y = 1$; (ii) its reciprocal decays away to zero in the tails, therefore, from a (rounded) maximum value of unity at $x = 0$; (iii) multiplying this by $x$, or a straight line through the origin, we see that $x/(1+x^2)$ rises to a maximum from $(0,0)$ before dying away (like $1/x$) to $y = 0$ as $x \to \infty$, and is the negative mirror image for $x < 0$.

## 2.5 Circles

The most perfect curve of all is probably a circle. Mathematically, it is defined as 'the locus of a point such that its distance from a fixed point is constant'. We can turn this formal jargon into an equation through a simple geometrical argument. Suppose that $(x_0, y_0)$ is the coordinate of the fixed point, called the *centre*, and that $(x, y)$ is some arbitrary point on the *circumference*. Then, by constructing a right-angled triangle and using *Pythagoras' theorem*, that 'the square of the hypotenuse is equal to the sum of the squares of the other two sides', we find that the distance from $(x_0, y_0)$ to $(x, y)$ is given by the square root of $(x - x_0)^2 + (y - y_0)^2$. Since this has to be a constant for all points on the rim, we must have

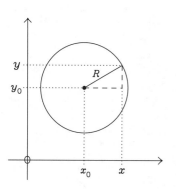

$$(x - x_0)^2 + (y - y_0)^2 = R^2 \tag{2.8}$$

where $R$ is the *radius* of the circle. If the centre happens to be at the origin, then eqn (2.8) reduces to $x^2 + y^2 = R^2$.

Equation (2.8) can easily be expanded and rearranged into a 'standard', if less obvious, form

$$x^2 + y^2 + 2\alpha x + 2\beta y + \gamma = 0 \tag{2.9}$$

where the centre is at $(-\alpha, -\beta)$, and the radius is given by the square root of $\alpha^2 + \beta^2 - \gamma$. For completeness, we should also add that the circumference of a circle is equal to $2\pi R$ (the definition of $\pi$) and that its area is $\pi R^2$.

## 2.6 Ellipses

If a circle is squashed so that it is shorter in one direction, and longer at right-angles to it, then an *ellipse* is obtained. The simplest form of the equation that describes this situation is

$$\frac{x^2}{a^2} + \frac{y^2}{b^2} = 1 \tag{2.10}$$

It corresponds to the case where the centre is at the origin, and the *principal axes* are along those of the $x$ and $y$ coordinates; the widths in these directions are given by $2a$ and $2b$. We recover the circle if $a = b$, of course, so that the denominators on the left-hand side are equal to $R^2$. If the centre was at $(x_0, y_0)$, then $x$ would be replaced by $(x-x_0)$ and $y$ by $(y-y_0)$ in eqn (2.10); we could also write a formula very similar to eqn (2.9), but the coefficients of the $x^2$ and $y^2$ terms would be different. If the *major* (or long) and *minor* (short) axes did not lie along the $x$ and $y$ coordinate directions, then there would be an additional $xy$ cross-term. While it can be shown that the area of the ellipse in eqn (2.10) is $\pi ab$, there is no simple formula for its perimeter.

Both the ellipse and the parabola are, in fact, linked through the traditional topic of *conic sections*. That is to say, if an ordinary cone is cut in various ways, then depending on the direction of the slice, the resulting cross-section is either an ellipse, or a parabola, or a *hyperbola*. Mathematically, all three correspond to the path of a point $P$, with coordinates $(x, y)$, which moves such that the ratio of its distance $r$ from a fixed point $S$ (called the *focus*) to its perpendicular distance $l$ from a given reference line (known as the *directrix*) is constant; so, defining the latter to be the *eccentricity* $\epsilon$, we have

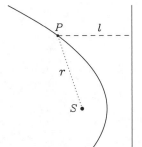

$$\frac{r}{l} = \epsilon$$

If $\epsilon < 1$, then we obtain an ellipse; if $\epsilon = 1$, we get a parabola; and $\epsilon > 1$ gives a hyperbola. In terms of eqn (2.10), $S$ is at $(a\epsilon, 0)$ and the directrix is the line $x = a/\epsilon$; the symmetry of the situation means that $(-a\epsilon, 0)$ and $x = -a/\epsilon$ would be an equally valid choice. It can be shown further that $a$, $b$, and $\epsilon$ obey the relationship $b^2 = a^2(1-\epsilon^2)$, so that a circle is the special case of $\epsilon = 0$. The orbital path of a planet is a physical example of an ellipse, with the sun at one focus, as was discovered by Kepler in 1609.

We note in passing that the equation of a hyperbola is very similar to eqn (2.10) except for a minus sign between the two terms on the left-hand side; it gives an open curve (two symmetrical ones actually), rather like a parabola, that asymptotically approaches the straight lines $y = \pm bx/a$ as $x$ becomes very large.

## 2.7 Worked examples

---

**(1)** Find the equation of the straight line that passes through the two points $(-1, 3)$ and $(3, 1)$; where does it intersect with $y = x + 1$?

---

General equation of a straight line is $y = mx + c$.

Passes through $(-1, 3) \Rightarrow 3 = -m + c$ $\left.\right\}$   $m = -1/2$

Passes through $\phantom{(-}(3, 1) \Rightarrow 1 = 3m + c$   $c = 5/2$

$\therefore$ Equation of line is $\underline{2y = 5 - x}$

For intersection,   $\left. \begin{matrix} 2y = 5 - x \\ y = 1 + x \end{matrix} \right\}$   $\begin{matrix} y = 2 \\ x = 1 \end{matrix}$

$\therefore$ Lines intersect at $\underline{(1, 2)}$

**(2)** By 'completing the square', find the coordinates of the turning point of $y = x^2 + x + 1$; hence sketch the parabola.

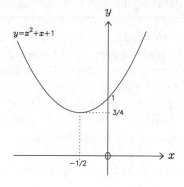

$$y = x^2 + x + 1 = \left(x + \tfrac{1}{2}\right)^2 - \tfrac{1}{4} + 1$$

$$= \left(x + \tfrac{1}{2}\right)^2 + \tfrac{3}{4}$$

Smallest value of $y$ when $x + 1/2 = 0$

$\therefore$  Turning point is at  $\underline{(-1/2, 3/4)}$

**(3)** Where does the curve $y = (x-3)(x-1)(x+1)$ cross the $x$ and $y$ axes? Hence sketch this cubic function, and state the ranges of $x$ values for which it is greater than zero.

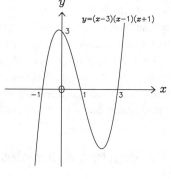

$y = 0$ when $x = 3$, $x = 1$, and $x = -1$.

When $x = 0$, $y = 3$.

$y > 0$  when  $|x| < 1$  or  $x > 3$

**(4)** Sketch the functions $y = 1 - e^{-x}$ and $y = 1 - e^{-2x}$ for positive values of $x$; and $y = e^{-|x|}$, $y = 1/x$, and $y = 1/(x^2 - 1)$ for all $x$.

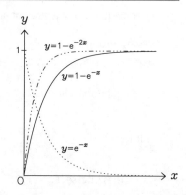

Since the exponential function $e^{-x}$ decays from a value of one at $x = 0$ to zero as $x \to \infty$, $1 - e^{-x}$ rises from the origin to an asymptotic limit of $y = 1$ as $x$ becomes very large; $e^{-2x}$ decays twice as quickly as $e^{-x}$, therefore $1 - e^{-2x}$ rises faster than $1 - e^{-x}$ to its ultimate value of $y = 1$.

With reference to chapter 4, neither $e^{-|x|}$ nor $1/x$ is differentiable at the origin. That is to say, because of their cuspy and discontinuous behaviour respectively, neither function has a well-defined gradient at $x=0$. Also, $e^{-|x|}$ is an *even* or *symmetric* function and $1/x$ is an *odd* or *antisymmetric* function; this refers to the symmetry of the curve with respect to its reflection given by an imagined mirror along $x=0$.

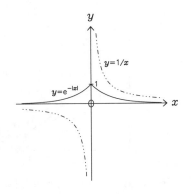

The curve $y=1/x$ is an example of a hyperbola, with the axes as asymptotes.

The best way of sketching a complicated function is often to decompose it into a series of more straightforward steps. For $1/(x^2-1)$, for example, first sketch the parabola $y=x^2-1$; wherever the magnitude of $y$ is large its reciprocal will be small, and vice versa, but of the same sign.

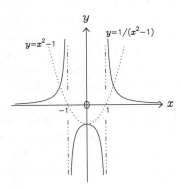

---

**(5)** Given the Arrhenius relationship between the rate constant $k$ and temperature $T$ of a chemical reaction, $k = A\,e^{-E_a/RT}$ where R is the gas constant, sketch a straight line graph that can be used to determine the activation energy $E_a$ and the pre-exponential factor A.

Taking the natural logarithm of both sides of the Arrhenius equation,

$$\ln(k) = \ln(A) - \left[\frac{E_a}{R}\right]\frac{1}{T}$$

This gives a straight line when $\ln(k)$ is plotted against $1/T$, with an intercept of $\ln(A)$ and a gradient of $-E_a/R$.

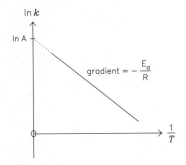

---

**(6)** The Maxwell-Boltzmann distribution of the kinetic energies, $E$, of particles in an ideal gas is given by $f(E) = \alpha\sqrt{E}\,e^{-E/kT}$ where k is the Boltzmann constant, T is the temperature and $\alpha = 2/\sqrt{\pi(kT)^3}$. Sketch $f(E)$ for two temperatures, where $T_1 > T_2$.

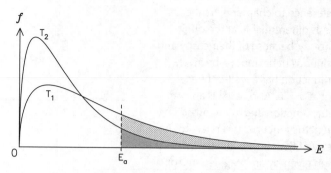

The Maxwell-Boltzmann distribution is a product of two functions: one proportional to $\sqrt{E}$, which increases with energy, and the other an exponential decay from unity. $f(E)$ therefore rises from the origin, initially proportional to $\sqrt{E}$, before reaching a maximum and decaying to zero. Since the exponential factor decays more slowly as the temperature rises, the location of the maximum increases to higher energies with T.

The proportionality constant $\alpha$ is chosen such that the total area under the curve is fixed to be unity. A distribution that is more spread out will consequently have a maximum of a lower amplitude. The proportion of gas particles that have an energy above an activation threshold $E_a$ is given by area under the curve for $E \geqslant E_a$. Hence, the extra fraction of molecules that reach the activation energy at $T_1$ over that at the lower temperature of $T_2$ is indicated by the hatched area.

# 3 Trigonometry

## 3.1 Angles and circular measure

An angle is a measure of rotation or turn, and is usually specified in *degrees*. There are 360° in a complete twist, where one ends up facing the same way as at the beginning; 90° therefore represents a right-angle, 180° an about-face, and so on. Despite our familiarity with degrees, a circular measure that is often more useful in a mathematical context is a *radian*. It is a dimensionless quantity defined as follows: if a radius of length $R$ is spun through an angle $\theta$ and generates an arc of length $L$, then

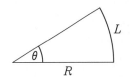

$$\theta = \frac{L}{R} \qquad (3.1)$$

Since the circumference of a circle is $2\pi R$, $360° = 2\pi$ radians; a right-angle is $\pi/2$ and, in general, an angle in degrees can be converted into one in radians by multiplying it by $\pi/180$. In other words, a radian is about $57.3°$.

Unless degrees are mentioned explicitly, it is best to assume that all angles are implicitly given in radians (especially if they contain factors of $\pi$).

## 3.2 Sines, cosines, and tangents

The most elementary definition of sines, cosines, and tangents is provided by the ratios of the sides of a right-angled triangle. That is to say, if the length of the hypotenuse is $r$, and that of the side adjacent and opposite to the angle $\theta$ is $x$ and $y$ respectively, then

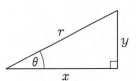

$$\sin \theta = \frac{y}{r}, \quad \cos \theta = \frac{x}{r}, \quad \tan \theta = \frac{y}{x} \qquad (3.2)$$

From this, it is easy to see that the three trigonometric quantities are linked through the relationship

$$\tan \theta = \frac{\sin \theta}{\cos \theta} \qquad (3.3)$$

Furthermore, as the angle between $y$ and $r$ is $90° - \theta$, because the sum around a triangle is $180°$, the sines and cosines of (acute) angles are complementary

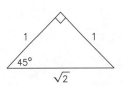

$$\sin \theta = \cos (\pi/2 - \theta) \qquad (3.4)$$

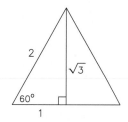

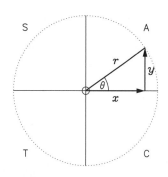

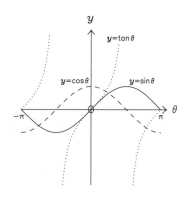

If $\theta$ is extremely small, then $y$ must be tiny while $r$ and $x$ are of virtually the same size; as such, eqns (3.2) and (3.4) tell us that $\sin(0) = \cos(\pi/2) = 0$ and $\cos(0) = \sin(\pi/2) = 1$. With some appropriate right-angled triangles, and Pythagoras' theorem, it can also be shown that $\sin(\pi/6) = \cos(\pi/3) = 1/2$, $\sin(\pi/4) = \cos(\pi/4) = 1/\sqrt{2}$ and $\sin(\pi/3) = \cos(\pi/6) = \sqrt{3}/2$. In conjunction with eqn (3.3), therefore, we get $\tan(0) = 0$, $\tan(\pi/6) = 1/\sqrt{3}$, $\tan(\pi/4) = 1$, $\tan(\pi/3) = \sqrt{3}$, and $\tan(\pi/2) \to \infty$.

Our discussion has so far been restricted to the case of $0° \leqslant \theta \leqslant 90°$. It can be extended to any angle by generalizing the definitions in eqn (3.2) so that $x$ and $y$ are the coordinates of a point that is a distance $r$ from the origin and at angle $\theta$ from the positive $x$-axis when measured anticlockwise; a negative value of $\theta$ represents a clockwise rotation. Then we find that the sine, cosine, and tangent of an arbitrary angle can always be related to one that lies between $0°$ and $90°$, apart from a possible minus sign. To see this, the first thing to notice is that the trigonometric functions are *periodic* in that they identically repeat themselves every $360°$

$$\sin\theta = \sin(\theta + 2\pi N) \tag{3.5}$$

where N is an integer, and similarly for $\cos\theta$ and $\tan\theta$. If $90° < \theta \leqslant 180°$, so that $x$ is negative and $y$ is positive, then $\tan\theta = y/x$ is equal to minus the 'opposite over the adjacent' of a right-angled triangle with the supplementary angle of $180° - \theta$; in other words, $\tan\theta = -\tan(\pi - \theta)$. The ratios of $x$ and $y$ with respect to the hypotenuse $r$ yield $\sin\theta = \sin(\pi - \theta)$ and $\cos\theta = -\cos(\pi - \theta)$. A repetition of this argument for the cases $180° < \theta \leqslant 270°$ and $270° < \theta \leqslant 360°$ leads to the general result that the sine, cosine, and tangent of $\theta$ are equal to the value of the trigonometric function for the smallest corresponding angle between the 'radius' and the $x$-axis, but with a sign dependent on the quadrant. The *parity* of the latter can be remembered with the acronym CAST, which is made up from the first letters of 'cosine, all, sine, and tangent', as they are placed anticlockwise around the origin starting from the bottom right-hand corner, and refers to the functions that are positive in that quarter. For example, $\tan(210°) = \tan(30°)$, and $\tan(300°) = -\tan(60°)$.

Before leaving this section, we should make two further points. The first is simply a matter of nomenclature

$$\sec\theta = \frac{1}{\cos\theta}, \quad \operatorname{cosec}\theta = \frac{1}{\sin\theta}, \quad \cot\theta = \frac{1}{\tan\theta} \tag{3.6}$$

where the abbreviations stand for *secant*, *cosecant* and *cotangent*. The second concerns the approximation of $\sin\theta$, $\cos\theta$, and $\tan\theta$ when the angle is small but finite. We saw earlier that $\sin(0) = 0$, $\cos(0) = 1$ and $\tan(0) = 0$, but a more careful consideration of how the sector in the definition of circular measure starts to resemble a right-angled triangle as $\theta$ tends to zero leads to

$$\sin\theta \approx \theta, \quad \tan\theta \approx \theta, \quad \cos\theta \approx 1 - \theta^2/2 \tag{3.7}$$

where $|\theta| \ll 1$, and is specified in radians. It can also be verified by looking at the plots of the trigonometric functions in the neighbourhood of $\theta = 0$.

## 3.3 Pythagorean identities

One of the most basic properties of a right-angled triangle that we learn about is Pythagoras' theorem; in our present set-up, it takes the form $x^2 + y^2 = r^2$. If we divided both sides by $r^2$, so that $(x/r)^2 + (y/r)^2 = 1$, then substitution from eqn (3.2) would yield the relationship

$$\sin^2\theta + \cos^2\theta = 1 \tag{3.8}$$

where we have followed the convention that $\sin^2\theta = (\sin\theta)^2$, and so on. This formula, along with many others in this chapter, is often written with a three-pronged equals sign ($\equiv$) to indicate that it is an *identity*; this differs from an ordinary equation (such as $\sin\theta = \cos\theta$) in that it holds for all values of $\theta$, rather than just a few specific ones. Similar divisions of Pythagoras' theorem by $x$ and $y$, instead of $r$, give rise to two additional identities

$$\tan^2\theta + 1 = \sec^2\theta \quad \text{and} \quad \cot^2\theta + 1 = \csc^2\theta \tag{3.9}$$

Incidentally, eqn (3.8) yields eqn (2.10) if $x = a\cos\theta$ and $y = b\sin\theta$; this is known as the *parametric* form of an ellipse. The equivalent formulation for a circle (centred at the origin) is $x = R\cos\theta$ and $y = R\sin\theta$.

## 3.4 Compound angles

If an angle is expressed as the sum of two others, such as $\theta = A + B$, then its sine can be written in terms of the trigonometric functions of its constituent parts. While the derivation of the formula requires a bit of geometrical and algebraic inspiration, the result is easy to state

$$\sin(A + B) = \sin A \cos B + \cos A \sin B \tag{3.10}$$

A special case occurs when $A = B$ with the sine of a double-angle reducing to

$$\sin 2A = 2\sin A \cos A \tag{3.11}$$

The sine of a difference between two angles, say $A - B$, can be obtained from eqn (3.10) by noting that $\sin(-\theta) = -\sin\theta$ and $\cos(-\theta) = \cos\theta$

$$\sin(A - B) = \sin A \cos B - \cos A \sin B \tag{3.12}$$

The same geometrical construction, and similar algebraic reasoning, leads to the following formulae for the cosine of the sum and difference $A \pm B$

$$\cos(A \pm B) = \cos A \cos B \mp \sin A \sin B \tag{3.13}$$

where the negative sign on the right-hand side goes with the positive one on the left, and vice versa. If $A = B$, then the cosine of a double-angle reduces to

$$\cos 2A = \cos^2 A - \sin^2 A \tag{3.14}$$

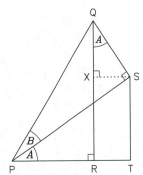

$$\sin(A + B) = \frac{QR}{PQ}$$
$$= \frac{QX + ST}{PQ}$$
$$= \frac{QS}{PQ}\cos A + \frac{PS}{PQ}\sin A$$
$$= \sin B \cos A + \cos B \sin A$$

With a substitution from eqn (3.8), it can also be written in two alternative forms that either contain only $\sin A$ or just $\cos A$

$$\cos 2A = 2\cos^2 A - 1 = 1 - 2\sin^2 A \tag{3.15}$$

According to eqn (3.3), the divisions of eqns (3.10) and (3.12) by their counterparts in eqn (3.13) give the formulae for the tangent of the sum and difference of two angles

$$\tan(A \pm B) = \frac{\tan A \pm \tan B}{1 \mp \tan A \tan B} \tag{3.16}$$

The result $\tan 2A = 2\tan A/(1-\tan^2 A)$ then follows from putting $A = B$.

## 3.5 Factor formulae

Useful identities, relating the sums and differences of sines and cosines to their products, can be derived from the formulae of the previous section. For example, the addition of eqns (3.10) and (3.12) gives

$$\sin(A+B) + \sin(A-B) = 2\sin A \cos B \tag{3.17}$$

The simple substitution $X = A+B$ and $Y = A-B$ then yields

$$\sin X + \sin Y = 2\sin\left(\tfrac{X+Y}{2}\right)\cos\left(\tfrac{X-Y}{2}\right) \tag{3.18}$$

$$\sin X - \sin Y = 2\cos\left(\tfrac{X+Y}{2}\right)\sin\left(\tfrac{X-Y}{2}\right)$$

Although these equations are equivalent, the former is better for decomposing a product into a sum and the latter is more suitable for going the other way. The subtraction of eqns (3.10) and (3.12) leads to expressions similar to eqns (3.17) and (3.18), except that there is a minus sign on the left-hand side and an interchange between the sin and cos on the right.

A corresponding manipulation for the cosines of compound angles, from eqn (3.13), leads to

$$\cos(A+B) + \cos(A-B) = 2\cos A \cos B \tag{3.19}$$

and

$$\cos(A+B) - \cos(A-B) = -2\sin A \sin B \tag{3.20}$$

$$\cos X - \cos Y = -2\sin\left(\tfrac{X+Y}{2}\right)\sin\left(\tfrac{X-Y}{2}\right)$$

along with their $X$ and $Y$ counterparts.

## 3.6 Inverse trigonometric functions

If we knew that the sine of an angle $\theta$ in a right-angled triangle was equal to a half, then we could deduce that $\theta$ was $30°$. This kind of reverse operation is an example of using an *inverse* trigonometric function. That is to say, the case above could be generalized with the formal definition

$$y = \sin\theta \iff \theta = \sin^{-1} y \tag{3.21}$$

We should caution that the 'power of minus one' notation here is anomalous, though widespread, in that it does not mean the reciprocal of $\sin y$; a less confusing, but rarer, alternative for the inverse function is arcsin. Expressions very similar to eqn (3.21) can be written for all the trigonometric functions, where $\arccos y = \cos^{-1} y$, $\arctan y = \tan^{-1} y$, and so on.

To appreciate the nature of the relationship $\theta = \sin^{-1} y$ we simply need to turn the plot of $y = \sin\theta$ through $90°$, so that the $y$-axis is horizontal and the $\theta$-axis is vertical; the same is true for $\theta = \cos^{-1} y$ and $\theta = \tan^{-1} y$ with respect to the graphs of $\cos\theta$ and $\tan\theta$. It then follows that the inverse trigonometric functions are *multivalued*, since a given sin, cos, or tan can be obtained from more than one angle. If we were only told that $\sin\theta = 1/2$, for example, then $\theta$ could be $30°$ or $150°$; or, indeed, any multiple number of $360°$ in addition to these two values. The other main point to note is that $\sin^{-1} y$ and $\cos^{-1} y$ do not exist if the magnitude of $y$ is greater than unity, because $|\sin\theta| \leqslant 1$ and $|\cos\theta| \leqslant 1$ for all $\theta$; there is no such restriction for $\tan^{-1} y$, of course, as $\tan\theta$ can lie anywhere in the range $\pm\infty$.

$$\cos^{-1}\left(\tfrac{1}{2}\right) = \pm\tfrac{\pi}{3} + 2n\pi,$$

$$\tan^{-1}\left(\sqrt{3}\right) = \tfrac{\pi}{3} + n\pi,$$

$$n = 0, \pm 1, \pm 2, \ldots$$

## 3.7 The sine and cosine rules

Let us conclude this chapter with a statement of the sine and cosine formulae, which relate the lengths of the sides of a general triangle to its angles; their proofs will become straightforward after dealing with the topic of *vectors*. If the angles of a triangle are labeled as $A$, $B$, and $C$, then the convention is to denote the sides opposite them with the corresponding lowercase letters $a$, $b$, and $c$. With this set-up, it can be shown that

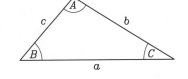

$$\frac{a}{\sin A} = \frac{b}{\sin B} = \frac{c}{\sin C} \tag{3.22}$$

and

$$a^2 = b^2 + c^2 - 2bc \cos A \tag{3.23}$$

These are known as the sine and cosine rules , respectively. The former can be split up into three separate relationships, since this is the number of equal pairs in eqn (3.22). The latter also yields three equations, because the sides and angles in eqn (3.23) can be permuted; e.g. $b^2 = c^2 + a^2 - 2ca \cos B$.

## 3.8 Worked examples

(1) Solve the following in the range $-\pi$ to $\pi$:
  (a) $\tan\theta = -\sqrt{3}$, (b) $\sin 3\theta = -1$ and (c) $4\cos^3\theta = \cos\theta$.

(a) $\tan\theta = -\sqrt{3} \quad \Rightarrow \quad \theta = -\tfrac{\pi}{3}$ or $\tfrac{2\pi}{3}$

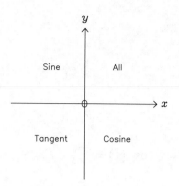

Sine | All

Tangent | Cosine

(b) $\sin 3\theta = -1 \quad \Rightarrow \quad 3\theta = -\frac{5\pi}{2}$ or $-\frac{\pi}{2}$ or $\frac{3\pi}{2}$

$$\therefore \quad \underline{\theta = -\frac{5\pi}{6} \text{ or } -\frac{\pi}{6} \text{ or } \frac{\pi}{2}}$$

(c) $4\cos^3\theta - \cos\theta = 0 \quad \Rightarrow \quad \cos\theta\,(4\cos^2\theta - 1) = 0$

$$\therefore \quad \cos\theta = 0 \quad \text{or} \quad \cos\theta = \pm\tfrac{1}{2}$$

$$\therefore \quad \underline{\theta = \pm\tfrac{\pi}{2} \quad \text{or} \quad \theta = \pm\tfrac{\pi}{3} \text{ or } \pm\tfrac{2\pi}{3}}$$

---

**(2)** From the definitions of a radian and a sine, indicate why $\sin\theta \approx \theta$ for small angles; show how this leads to $\cos\theta \approx 1 - \theta^2/2$.

$$\sin\theta = \frac{AB}{OB} \quad \text{and} \quad \theta = \frac{\text{Arc-length } AC}{OC}$$

As $\theta \to 0$, $\quad AB \to \text{Arc-length } AC$

$$OB \to OC$$

$$\therefore \quad \sin\theta \to \theta$$

i.e. $\quad \underline{\sin\theta \approx \theta} \quad$ for $\theta \ll 1$

Also, "$\cos 2\phi = 1 - 2\sin^2\phi$"

$$\therefore \quad \cos\theta \approx 1 - 2\left(\tfrac{\theta}{2}\right)^2 \qquad \text{i.e.} \quad \underline{\cos\theta \approx 1 - \tfrac{\theta^2}{2}} \quad \text{for } \theta \ll 1$$

---

**(3)** If $t = \tan(\theta/2)$, express $\tan\theta$, $\cos\theta$, and $\sin\theta$ in terms of $t$.

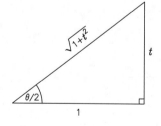

If $\tan(\theta/2) = t$,

Pythagoras' $\Rightarrow \quad \sin(\theta/2) = \dfrac{t}{\sqrt{1+t^2}} \quad$ and $\quad \cos(\theta/2) = \dfrac{1}{\sqrt{1+t^2}}$

But $\sin\theta = 2\sin(\theta/2)\cos(\theta/2) \qquad \therefore \quad \underline{\sin\theta = \dfrac{2t}{1+t^2}}$

$$\cos\theta = \cos^2(\theta/2) - \sin^2(\theta/2) \qquad \therefore \quad \underline{\cos\theta = \dfrac{1-t^2}{1+t^2}}$$

$$\tan\theta = \frac{\sin\theta}{\cos\theta} \qquad \therefore \quad \underline{\tan\theta = \dfrac{2t}{1-t^2}}$$

**(4)** Show that $a \sin\theta + b \cos\theta$ can be written as $A \sin(\theta + \phi)$, where $A$ and $\phi$ are related to $a$ and $b$. Hence solve $\sin\theta + \cos\theta = \sqrt{3/2}$.

If $\quad a \sin\theta + b \cos\theta = A \sin(\theta + \phi)$

$$= A \cos\phi \, \sin\theta + A \sin\phi \, \cos\theta$$

Then $\left. \begin{array}{l} a = A \cos\phi \\[2mm] b = A \sin\phi \end{array} \right\}$ $\qquad \dfrac{b}{a} = \dfrac{\sin\phi}{\cos\phi} = \tan\phi$

$$a^2 + b^2 = A^2(\sin^2\phi + \cos^2\phi) = A^2$$

i.e. $\quad \underline{A = \sqrt{a^2 + b^2} \quad \text{and} \quad \phi = \tan^{-1}(b/a)}$

$$\sin\theta + \cos\theta = \sqrt{\tfrac{3}{2}} \quad \Rightarrow \quad \sqrt{2}\,\sin\left(\theta + \tfrac{\pi}{4}\right) = \sqrt{\tfrac{3}{2}}$$

$$\therefore \quad \sin\left(\theta + \tfrac{\pi}{4}\right) = \tfrac{\sqrt{3}}{2}$$

$$\therefore \quad \theta + \tfrac{\pi}{4} = \tfrac{\pi}{3} \text{ or } \tfrac{2\pi}{3}$$

$$\text{i.e.} \quad \theta = \tfrac{\pi}{12} \text{ or } \tfrac{5\pi}{12}$$

**(5)** Show that $8 \sin^4\theta = \cos 4\theta - 4 \cos 2\theta + 3$ and find a similar expression for $\cos^4\theta$.

$$\cos 2\theta = 1 - 2\sin^2\theta = 2\cos^2\theta - 1$$

$$\therefore \quad 8 \sin^4\theta = 8\left(\sin^2\theta\right)^2 = 8\left(\frac{1 - \cos 2\theta}{2}\right)^2$$

$$= 2\left(1 - 2\cos 2\theta + \cos^2 2\theta\right)$$

$$= 2 - 4\cos 2\theta + \left(2\cos^2 2\theta - 1\right) + 1$$

$$= \underline{3 - 4\cos 2\theta + \cos 4\theta}$$

$$\therefore \quad 8 \cos^4\theta = 8\left(\cos^2\theta\right)^2 = 8\left(\frac{\cos 2\theta + 1}{2}\right)^2$$

$$= 2\left(\cos^2 2\theta + 2\cos 2\theta + 1\right)$$

$$= \left(2\cos^2 2\theta - 1\right) + 1 + 4\cos 2\theta + 2$$

$$= \underline{\cos 4\theta + 4\cos 2\theta + 3}$$

**(6)** A triatomic molecule has bond-lengths of 1.327 Å and 1.514 Å , and a bond-angle of 107.5°; find the distance between the furthest atoms.

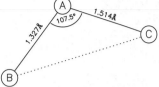

Cosine rule: $BC^2 = AB^2 + AC^2 - 2\,AB.AC\cos(B\hat{A}C)$

$\therefore$ Distance$^2 = 1.327^2 + 1.514^2 - 2 \times 1.327 \times 1.514 \times \cos(107.5°)$

$= 5.2614$ Å$^2$

$\therefore$ Distance between furthest atoms $= \underline{2.294\text{ Å}}$

**(7)** $PF_5$ is a trigonal bipyramidal molecule with a phosphorus at the centre. Three of the fluorines are equally spaced in a plane, with PF bond-lengths of 153 pm; the other two lie along a perpendicular axis, on either side of the plane, with PF bond-lengths of 158 pm. Calculate the distances between adjacent fluorine atoms.

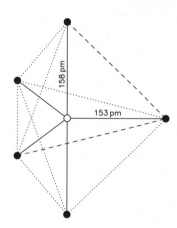

Distance between fluorine atoms in the plane $= 2 \times 153 \sin(60°)$

$= 153\sqrt{3} = \underline{265\text{ pm}}$

Distance between axial and planar fluorine atoms $= \sqrt{153^2 + 158^2}$

$= \underline{220\text{ pm}}$

# 4 Differentiation

## 4.1 Gradients and derivatives

In chapter 2 we saw how the relationship between two quantities, called $x$ and $y$, could be visualized with the aid of a graph. While the intersections of the curve with the $x$ and $y$ axes may be of interest, it is often more important to know the slope at any given point; that is, how quickly $y$ increases, or decreases, as $x$ changes, and vice versa. This issue is at the heart of the topic of *differentiation*, and most of this chapter is devoted to learning how to calculate the gradient algebraically.

Let us begin with a precise definition of what is meant by the slope of a curve. Suppose that $y$ is related to $x$ through some function called 'f', usually written as $y = f(x)$, so that $f(x) = mx + c$ for a straight line, $f(x) = \sin(x)$ for a sinusoidal variation and so on. Then, if the horizontal coordinate changes from $x$ by a very small amount, $\delta x$, to $x + \delta x$, the value of $y$ is altered from $f(x)$ to $f(x + \delta x)$. The gradient, at a point $x$, is defined to be the ratio of the change in the vertical coordinate, $\delta y$, to that of the horizontal increment, as $\delta x$ becomes infinitesimally small. This is written formally as

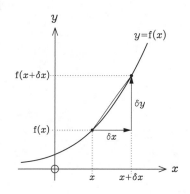

$$\frac{dy}{dx} = \lim_{\delta x \to 0} \frac{\delta y}{\delta x} = \lim_{\delta x \to 0} \frac{f(x + \delta x) - f(x)}{\delta x} \qquad (4.1)$$

where $dy/dx$ is known as the *derivative*, or *differential coefficient*, and is pronounced 'dy-by-dx'; it can also be denoted with a 'dashed' notation, $y'$ or $f'(x)$, or even a 'dotted' one, $\dot{y}$, if the $x$-axis pertains to time. The tendency of $\delta x \to 0$ has to be approached gradually to ascertain the *limiting* value of the ratio $\delta y/\delta x$, as both increments are equal to zero when the condition is met. Strictly speaking, we should check that the value of $dy/dx$ is independent of the sign of $\delta x$, but this is assured as long as the curve $y = f(x)$ is 'smooth'; problems will arise if kinks and sudden breaks (or *discontinuities*) are present, and the function is said to be non-differentiable at those points.

As a simple example of evaluating a derivative from 'first principles', let us consider the case of a straight line $y = mx + c$. Substituting for $f(x)$ into eqn (4.1) gives

$$\frac{dy}{dx} = \lim_{\delta x \to 0} \frac{m(x + \delta x) + c - (mx + c)}{\delta x} = \lim_{\delta x \to 0} \frac{m\,\delta x}{\delta x}$$

where the last term follows from elementary algebra. Although the numerator

$$y' = \frac{dy}{dx} = \frac{df}{dx} = f'(x)$$

$$\dot{y} = \frac{dy}{dt}$$

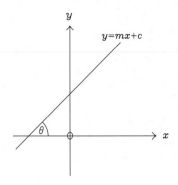

and denominator are both zero in the limit $\delta x \rightarrow 0$, their ratio is well-defined; indeed, as expected from the discussion in section 2.1, the derivative has the same value everywhere and is equal to the gradient $m$. Although we have specifically dealt with a straight line, there are a couple of points that apply more generally. The first is that $y$ increases with $x$ if $dy/dx > 0$, and it decreases if $dy/dx < 0$; the rate of the variation is given by the magnitude of the derivative. The second point is that

$$\frac{dy}{dx} = \tan \theta \tag{4.2}$$

where $\theta$ is the anticlockwise angle between the slope and the positive $x$-axis. The properties of a tangent also confirm the former observation, since $dy/dx$ has the same sign as $\theta$ and a magnitude which increases with the size of the angle.

Having illustrated differentiation with an elementary example, let us move on to a more interesting case: $y = \sin x$. The substitution of $f(x)$ into eqn (4.1) now gives

$$\frac{dy}{dx} = \lim_{\delta x \rightarrow 0} \frac{\sin(x + \delta x) - \sin x}{\delta x}$$

As always, the taking of the limit will be the very last step; first we must consider how the numerator behaves as $\delta x$ becomes tiny, but isn't actually zero. This can be done by expanding $\sin(x + \delta x)$ with eqn (3.10) and using the small-angle approximations for sine and cosine in eqn (3.7)

$$\sin(x + \delta x) \approx \left(1 - \frac{\delta x^2}{2}\right) \sin x + \delta x \cos x$$

where $x$ is measured in radians. Putting this into our definition for $dy/dx$, and simplifying the resultant expression, we obtain

$$\frac{dy}{dx} = \lim_{\delta x \rightarrow 0} \cos x - \frac{\delta x \sin x}{2}$$

Hence, on finally letting $\delta x \rightarrow 0$, we see that the derivative of $\sin x$ is $\cos x$. This can be verified from the plots of sine and cosine in section 3.2, although the $x$-axis there was labelled as $\theta$. At the origin, the slope of the $\sin x$ curve is inclined at $45°$; its derivative is therefore $\tan(\pi/4) = 1$, which is the same as $\cos(0)$. The gradient reduces gradually to zero between $0°$ and $90°$, as does the cosine. The sine curve then starts to dip downwards so that, just like $\cos x$, the slope becomes increasingly negative. And so on, for all values of $x$. A similar algebraic argument leads to the result that the derivative of $\cos x$ is $-\sin x$, and can also be verified graphically.

$$\frac{d}{dx}\left(\sin x\right) = \cos x$$

$$\frac{d}{dx}\left(\cos x\right) = -\sin x$$

## 4.2 Some basic properties of derivatives

One of the most fundamental properties of derivatives is their *linearity*. That is to say, they satisfy the rules

$$\frac{d}{dx}\left[A\, f(x)\right] = A\frac{df}{dx} \quad \text{and} \quad \frac{d}{dx}\left[f(x) + g(x)\right] = \frac{df}{dx} + \frac{dg}{dx} \tag{4.3}$$

where $A$ is a constant, $f(x)$ and $g(x)$ are functions of $x$, and $d/dx$ is the differential *operator* meaning 'the rate of change, with respect to $x$, of'. While the formal proof of eqn (4.3) follows from eqn (4.1), the essential point is that 'the derivative of a sum is equal to the sum of the derivatives'. This enables us to differentiate a polynomial, for example, by knowing that the derivative of $x^M$ is $Mx^{M-1}$

$$\frac{d}{dx}\left(a_0 + a_1 x + a_2 x^2 + \cdots + a_N x^N\right) = a_1 + 2a_2 x + \cdots + N a_N x^{N-1} \quad (4.4)$$

The differential coefficient of $x^M$ itself can be derived from eqn (4.1) by using the binomial expansion of $(x+\delta x)^M$, as in eqn (1.11), and noting that all the terms with powers of $\delta x$ greater than one are negligible. Although the result is only shown for positive integer values of M in this way, it turns out to hold for any power. Equation (4.4) also confirms our earlier finding that the derivative of a straight line, which corresponds to $N = 1$, is a constant ($a_1$); as expected, the gradient of the case with no variation, $y = a_0$, is zero.

$$\frac{d}{dx}\left(x^M\right) = M x^{M-1}$$

Another important property of $dy/dx$ is its reciprocal relationship with its counterpart $dx/dy$

$$\frac{dy}{dx} = \frac{1}{dx/dy} \quad (4.5)$$

While the former is the slope of a curve in an ordinary graph, the latter is the gradient when $y$ is plotted horizontally and $x$ vertically; in other words, $dx/dy$ is the rate of change of $x$ with respect to $y$. The proof of eqn (4.5) is almost self-evident, in that it's obviously true for arbitrarily small (but finite) increments $\delta x$ and $\delta y$.

A good example of the use of eqn (4.5) is provided by the derivatives of inverse trigonometric functions. Following eqn (3.21), we have

$$y = \sin^{-1} x \quad \Longleftrightarrow \quad x = \sin y$$

The differentiation of the right-hand expression with respect to $y$ gives

$$\frac{dx}{dy} = \cos y = \sqrt{1 - \sin^2 y} = \sqrt{1 - x^2}$$

where we have used the identity of eqn (3.8) to relate $\cos y$ to $\sin y$, and hence write $dx/dy$ in terms of $x$; we have implicitly assumed that $|x| \leqslant 1$ and $|y| \leqslant \pi/2$. According to eqn (4.5), therefore, the derivative of arcsin $x$ is equal to the reciprocal of the square root of $1 - x^2$. A similar analysis for arccos $x$ shows that its derivative is minus that for arcsin $x$.

$$\frac{d}{dx}\left(\sin^{-1} x\right) = \frac{1}{\sqrt{1 - x^2}}$$

$$\frac{d}{dx}\left(\cos^{-1} x\right) = \frac{-1}{\sqrt{1 - x^2}}$$

The final point in this section is about the differentiation of derivatives. If, like $y = f(x)$, $dy/dx = f'(x)$ is a 'smooth' function of $x$, then it too can be differentiated to yield the *second derivative*

$$y'' = \frac{d^2 y}{dx^2} = \frac{d}{dx}\left(\frac{dy}{dx}\right) = \lim_{\delta x \to 0} \frac{f'(x+\delta x) - f'(x)}{\delta x} = f''(x) \quad (4.6)$$

Rather than telling us how $y$ changes with $x$, it conveys how the slope of $y$ varies with $x$. If $y$ represents *distance* travelled and $x$ time, for example, then $y'$ (or $\dot{y}$)

$$\frac{\mathrm{d}^n y}{\mathrm{d}x^n} = \frac{\mathrm{d}}{\mathrm{d}x}\left(\frac{\mathrm{d}^{n-1}y}{\mathrm{d}x^{n-1}}\right)$$

gives the *speed* and $y''$ (or $\ddot{y}$) the *acceleration*. Eqn (4.6) generalizes to higher order derivatives, so that the third one, $\mathrm{d}^3 y/\mathrm{d}x^3$, is the rate of change of $\mathrm{d}^2 y/\mathrm{d}x^2$ with respect to $x$; and so on.

## 4.3 Exponentials and logarithms

In section 2.4, we encountered the exponential function $y = \exp(x)$. This has the property that its gradient varies with $x$ in the same way as the function itself

$$\frac{\mathrm{d}}{\mathrm{d}x}\left(e^x\right) = e^x \tag{4.7}$$

$e = 2.7182818285$

This can even be regarded as the definition of the number 'e'. The derivative of the natural logarithm, $y = \ln(x)$, then follows from the differentiation of the equivalent exponential expression, $x = \exp(y)$, with respect to $y$

$$\frac{\mathrm{d}x}{\mathrm{d}y} = e^y = x$$

Hence, with the reciprocal relationship of eqn (4.5), we have

$$\frac{\mathrm{d}}{\mathrm{d}x}\left[\ln(x)\right] = \frac{1}{x} \tag{4.8}$$

$$\frac{\mathrm{d}}{\mathrm{d}x}\left(a^x\right) = a^x \ln(a)$$

$$\frac{\mathrm{d}}{\mathrm{d}x}\left[\log_a(x)\right] = \frac{1}{x \ln(a)}$$

The derivative of a logarithm, or a power, to any other base, say $a$, can be obtained by using the results of section 1.2 in conjunction with those above. For example, eqn (1.8) allows us to write $\log_a(x) = \ln(x)/\ln(a)$; thus the derivative of $\log_a(x)$ is equal to that of $\ln(x)$ divided by $\ln(a)$. Similarly, the differential coefficient of $a^x$ turns out to be $a^x$ times $\ln(a)$.

## 4.4 Products and quotients

The linearity of eqn (4.3) tells us how to differentiate the sum $y = u(x) + v(x)$, where u and v are functions of $x$, but what about the derivative of a product, $y = u(x)\,v(x)$? The answer can be derived from eqn (4.1), as long as we note that $u(x+\delta x)$ and $v(x+\delta x)$ may be expressed as $u + \delta u$ and $v + \delta v$, respectively, where $\delta u$ and $\delta v$ are the small changes in the values of u and v generated by the increment $\delta x$. Then, by the definition of a derivative, we have

$$\frac{\mathrm{d}y}{\mathrm{d}x} = \lim_{\delta x \to 0} \frac{(u+\delta u)(v+\delta v) - u\,v}{\delta x} = \lim_{\delta x \to 0} u\frac{\delta v}{\delta x} + v\frac{\delta u}{\delta x} + \frac{\delta u\,\delta v}{\delta x}$$

and on taking the limit $\delta x \to 0$, we obtain

$$\frac{\mathrm{d}}{\mathrm{d}x}(u\,v) = u\frac{\mathrm{d}v}{\mathrm{d}x} + v\frac{\mathrm{d}u}{\mathrm{d}x} \tag{4.9}$$

$(u\,v\,w)' = u\,v\,w' + u\,v'\,w + u'\,v\,w$

because $\delta u\,\delta v/\delta x \to 0$. This formula can be extended to the product of any number of terms, by either returning to first principles or by putting $v = f\,g$.

As an example of the use of eqn (4.9), consider the function $y = 2^x(x^3 - 1)$. This is equivalent to having $u = 2^x$ and $v = x^3 - 1$, so that their derivatives are $u' = 2^x \ln(2)$ and $v' = 3x^2$, whereby

$$\frac{d}{dx}\left[2^x(x^3 - 1)\right] = 2^x\left[3x^2 + (x^3 - 1)\ln(2)\right]$$

With practice, the explicit substitution of the functions $u$ and $v$ into eqn (4.9) is not required; it is simpler to implement it directly by remembering that the differential coefficient of a product is given by 'the first one times the derivative of the second plus the second times the derivative of the first'.

Incidentally, the repeated differentiation of a product of two terms leads to an expression analogous to the binomial expansion of eqn (1.11)

$$(uv)'' = uv'' + 2u'v' + u''v$$

$$\frac{d^n}{dx^n}(uv) = \sum_{r=0}^{n} {}^nC_r \frac{d^r u}{dx^r} \frac{d^{n-r}v}{dx^{n-r}} \qquad (4.10)$$

This is known as Leibnitz' theorem.

The derivative of a *quotient* or ratio, $y = u(x)/v(x)$, may be ascertained by applying the product rule to $u = vy$; the resulting equation, $u' = vy' + yv'$, can then be manipulated algebraically to yield $y'$

$$\frac{d}{dx}\left[\frac{u}{v}\right] = \frac{vu' - uv'}{v^2} \qquad (4.11)$$

A straightforward example of the use of this formula is in the calculation of the differential coefficient of $\tan x$:

$$\frac{d}{dx}(\tan x) = \frac{d}{dx}\left(\frac{\sin x}{\cos x}\right) = \frac{\cos^2 x + \sin^2 x}{\cos^2 x}$$

$$\frac{d}{dx}(\tan x) = \sec^2 x$$

where we have assumed a knowledge of the derivatives of $\sin x$ and $\cos x$, as well as eqns (3.3) and (4.11). Since the numerator on the far right is unity, by eqn (3.8), the differential coefficient of $\tan x$ is equal to one over $\cos^2 x$; or, with eqn (3.6), just $\sec^2 x$. This analysis can be retraced to show that the derivative of $\cot x$ is $-\text{cosec}^2 x$, or the result for the tangent used directly to differentiate its inverse form $y = \tan^{-1}x$ (as for the arcsin earlier); the latter turns out to be given by the reciprocal of $1 + x^2$.

$$\frac{d}{dx}\left(\tan^{-1}x\right) = \frac{1}{1 + x^2}$$

$$\frac{d}{dx}(\cot x) = -\text{cosec}^2 x$$

## 4.5 Functions of functions

We have now seen how to differentiate power-laws, exponentials, logarithms, trigonometric functions, and arithmetical combinations of them. Frequently, however, we encounter these familiar entities in more complicated settings; for example, $y = \ln(2 + \cos x)$, or $y = A\sin^2(\omega x + \phi)\exp(-kx)$ where all but $x$ and $y$ are constant. How can we use our knowledge about the derivatives of the underlying 'building blocks' to deal with the general situation?

Well, we need to enlist the help of a very powerful result known as the *chain rule*. This states that if $y$ is a function of u, and u is itself dependent on $x$, then $dy/dx$ is given by

$$\frac{dy}{dx} = \frac{dy}{du} \times \frac{du}{dx}$$

(4.12)

$$\frac{d}{dx}\left(\sec x\right) = \sec x \tan x$$

$$\frac{d}{dx}\left(\operatorname{cosec} x\right) = -\operatorname{cosec} x \cot x$$

Thus in our first example, $y = \ln(u)$ and $u = 2 + \cos x$. Since $dy/du = 1/u$ and $du/dx = -\sin x$, eqn (4.12) yields $dy/dx = -\sin x/(2 + \cos x)$. Several other straightforward cases, such as $y = \sec x$ and $y = \operatorname{cosec} x$, fall under the reciprocal category of $y = u^{-1}$; their derivatives, therefore, share the common form $dy/dx = -u^{-2}\,du/dx = -u'/u^2$. Indeed, the quotient rule of eqn (4.11) can be derived from eqn (4.9) by considering u/v to be the product of u and 1/v.

The proof of eqn (4.12) is, again, almost self-evident, in that it holds for arbitrarily small increments $\delta x$, $\delta y$, and $\delta u$ through simple division. In fact this indicates that the chain rule can be extended, as required, to deal with nested sets of functions. If $y = \sin[\ln(2 + \cos x)]$, for example, then we have $y = \sin u$, where $u = \ln(v)$ and $v = 2 + \cos x$; thus, by putting the 'standard' derivatives, like $dy/du = \cos u$, into

$$\frac{dy}{dx} = \frac{dy}{du} \times \frac{du}{dv} \times \frac{dv}{dx}$$

we obtain $dy/dx = -\cos[\ln(2 + \cos x)]\sin x/(2 + \cos x)$. With experience, the explicit substitution of u, v, ..., becomes unnecessary; familiarity eventually allows the chain rule to be implemented mentally in a stepwise manner.

Several of the differentiation rules that we have learnt may be needed for any given problem, and sometimes more than once. If $y = x \ln[x(2 + \cos x)]$, for example, then we have to deal with the product of $x$ and $\ln(u)$, where u is itself a product of $x$ and $2 + \cos x$. Thus $dy/dx = x\,u^{-1}\,du/dx + \ln u$, where $du/dx = -x \sin x + 2 + \cos x$.

As a final note in this section, we should mention an important variant of the chain rule

$$\frac{dy}{dx} = \frac{dy/dt}{dx/dt}$$

(4.13)

which is useful for differentiating parametric equations. That is when both $x$ and $y$ are expressed as functions of a common variable, denoted by $t$ in eqn (4.13). An example was met in section 3.3, where an ellipse was seen to take the form $x = a \cos\theta$ and $y = b \sin\theta$; it follows that $dy/dx = -b \cot\theta/a$.

## 4.6 Maxima and minima

One of the main reasons for being interested in the slopes of curves is that the places where the gradient is zero are often of great physical significance. They can be found by solving the equation

$$\frac{dy}{dx} = 0 \qquad (4.14)$$

and are known as *stationary* points. As the name suggests, they are locations where the value of $y$ remains virtually unchanged even if $x$ varies a little bit; in other words, they represent positions of 'equilibrium'.

Apart from the cases where $y$ gradually approaches a constant value as $x$ tends to infinity, as in $y = \exp(-x)$, there are three types of situations when eqn (4.14) is satisfied: (i) a *maximum*, which is like the top of a hill, where $y$ decreases on both sides; (ii) a *minimum*, which is akin to the base of a valley, so that $y$ increases around the stationary point; and (iii) a point of *inflexion*, which is a flat region where $y$ goes up on one side and down on the other. The first two scenarios are collectively called ' turning points ', and can be distinguished by considering how the sign of the gradient $dy/dx$ changes as the slope passes through the horizontal position

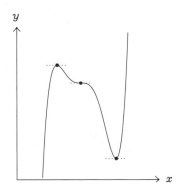

$$\frac{d^2y}{dx^2} < 0 \quad \text{for a max} \qquad \text{and} \qquad \frac{d^2y}{dx^2} > 0 \quad \text{for a min} \qquad (4.15)$$

Although the second derivative is zero at a point of inflexion, the condition that $d^2y/dx^2 = 0$ does not necessarily imply one. Indeed, this is a special case that always requires more careful thought. For example, it is easily shown that the curves $y = x^3$ and $y = x^4$ both have $dy/dx = 0$ and $d^2y/dx^2 = 0$ at the origin; from the graphs in section 2.3, however, it's clear that the quartic has a minimum while the cubic harbours a point of inflexion.

$$y = x^4$$
$$y' = 4\,x^3$$
$$y'' = 12\,x^2$$

As a concrete illustration of the discussion above, let's consider a hydrogen atom. *Quantum mechanics* shows that the probability that an electron in the 1$s$ (ground) state is at a distance $r$ from the nucleus is proportional to

$$p(r) = \frac{4\,r^2}{a_0^3}\,e^{-2r/a_0}$$

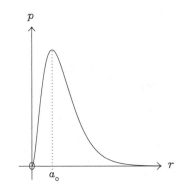

where the constant $a_0 = 0.5292 \times 10^{-10}$ m. The most likely radial position of the electron is then given by the maximum of $p(r)$, which is defined by the criterion of eqn (4.14)

$$\frac{dp}{dr} = \frac{8\,r}{a_0^3}\left(1 - \frac{r}{a_0}\right)e^{-2r/a_0} = 0$$

Of the three possible solutions, $r = 0$, $r = a_0$, and $r \to \infty$, only the middle one yields a negative value for the second derivative $d^2p/dr^2$; hence, according to eqn (4.15), the most probable location of the electron is at $r = a_0$.

There are many problems of physical interest where the minimum value of a function, such as the *Gibbs free energy*, or the mismatch of a model fit to experimental data, is of paramount importance. An example familiar from everyday experience is that of gravitational *potential energy*, where objects, be they marbles or bodies of water, settle naturally into the nearest hollow; as such, a minimum is known as a point of 'stable' equilibrium. By contrast, a maximum gives an 'unstable' equilibrium.

## 4.7 Implicit and logarithmic differentiation

So far, we have generally assumed that $y$ is an explicit function of $x$; often, however, the relationship between the two entities is sufficiently complicated that a rearrangement into the form $y = f(x)$ is not possible. Nevertheless, $dy/dx$ can still be ascertained, in terms of $x$ and $y$, by applying the operator $d/dx$ to both sides of the equation. If $x^2 y + \sin y = 6$, for example, then

$$\frac{d}{dx}\left(x^2 y\right) + \frac{d}{dx}\left(\sin y\right) = \frac{d}{dx}\left(6\right)$$

where we have used the linearity property of eqn (4.3) to split up the left-hand side into two parts. By the rules of differentiation for products, functions of functions, and so on, we therefore obtain

$$x^2 \frac{dy}{dx} + 2xy + \cos y \frac{dy}{dx} = 0$$

which can be manipulated algebraically to yield $dy/dx = -2xy/(x^2 + \cos y)$. This procedure, where an expression is differentiated 'as it stands', is usually called *implicit differentiation*.

Even when $y$ can be written as an explicit function of $x$, it is sometimes helpful to take its logarithm before differentiating it. This is particularly true for cases involving complicated products and quotients, such as

$$y = \frac{(x+5)\sqrt{(7+2x)^3}}{(2x^3+1)\cos x}$$

because, according to the rules of section 1.2, it turns the expression into a simpler one of sums and differences

$$\ln(y) = \ln(x+5) + \tfrac{3}{2}\ln(7+2x) - \ln(2x^3+1) - \ln(\cos x)$$

$$\frac{d}{dx}\left[\ln(y)\right] = \frac{dy}{dx} \times \frac{1}{y}$$

which is considerably easier to differentiate (implicitly), term by term. This logarithmic technique is also very useful for dealing with situations where $x$ occurs as a power. For example, if $y = (x^2+1)^x$ then an application of the differential operator $d/dx$ to both sides of $\ln(y) = x \ln(x^2+1)$, and a little algebra, yields $dy/dx = y\left[\ln(x^2+1) + 2x^2/(x^2+1)\right]$.

## 4.8 Worked examples

> **(1)** Differentiate the following from 'first principles':   (a) $y = x^n$, where $n$ is a positive integer; (b) $y = 1/x$; and (c) $y = \cos x$.

(a)   $\dfrac{dy}{dx} = \lim\limits_{\delta x \to 0} \dfrac{(x+\delta x)^n - x^n}{\delta x}$

$$= \lim_{\delta x \to 0} \frac{x^n + n\,x^{n-1}\delta x + \frac{1}{2}n(n-1)\,x^{n-2}\delta x^2 + \cdots + \delta x^n - x^n}{\delta x}$$

$$= \lim_{\delta x \to 0} \; nx^{n-1} + \tfrac{1}{2}n(n-1)\,x^{n-2}\delta x + \cdots + \delta x^{n-1}$$

$$\therefore \quad \frac{\mathrm{d}}{\mathrm{d}x}(x^n) = nx^{n-1}$$

(b) $\quad \dfrac{\mathrm{d}y}{\mathrm{d}x} = \lim_{\delta x \to 0} \dfrac{1/(x+\delta x) - 1/x}{\delta x}$

$$= \lim_{\delta x \to 0} \frac{x - (x+\delta x)}{(x+\delta x)\,x\,\delta x}$$

$$= \lim_{\delta x \to 0} \frac{-1}{(x+\delta x)\,x}$$

$$\therefore \quad \frac{\mathrm{d}}{\mathrm{d}x}(x^{-1}) = -x^{-2}$$

(c) $\quad \dfrac{\mathrm{d}y}{\mathrm{d}x} = \lim_{\delta x \to 0} \dfrac{\cos(x+\delta x) - \cos x}{\delta x}$

$$= \lim_{\delta x \to 0} \frac{\cos(\delta x)\cos x - \sin(\delta x)\sin x - \cos x}{\delta x}$$

$$= \lim_{\delta x \to 0} \frac{(1 - \frac{1}{2}\delta x^2)\cos x - \delta x \sin x - \cos x}{\delta x}$$

$$= \lim_{\delta x \to 0} \; -\tfrac{1}{2}\delta x \cos x - \sin x$$

$$\therefore \quad \frac{\mathrm{d}}{\mathrm{d}x}(\cos x) = -\sin x$$

---

**(2)** Differentiate the following functions, $y = f(x)$, with respect to $x$.

(a) $\quad y = (2x+1)^3 \quad \Rightarrow \quad \dfrac{\mathrm{d}y}{\mathrm{d}x} = 3(2x+1)^2 \times 2$

$$= 6(2x+1)^2$$

In this question, we have implicitly made the substitution $u = 2x+1$, so that $y = u^3$, and used the chain rule $\mathrm{d}y/\mathrm{d}x = \mathrm{d}y/\mathrm{d}u \times \mathrm{d}u/\mathrm{d}x$; the above result then follows from $\mathrm{d}y/\mathrm{d}u = 3u^2$ and $\mathrm{d}u/\mathrm{d}x = 2$.

(b)  $y = \sqrt{3x-1} = (3x-1)^{1/2}$  $\Rightarrow$  $\dfrac{dy}{dx} = \tfrac{1}{2}(3x-1)^{-1/2} \times 3$

$$= \frac{3}{2\sqrt{3x-1}}$$

(c)  $y = \cos(5x)$  $\Rightarrow$  $\dfrac{dy}{dx} = -\sin(5x) \times 5 = \underline{-5\sin(5x)}$

(d)  $y = \sin(3x^2+7)$  $\Rightarrow$  $\dfrac{dy}{dx} = \cos(3x^2+7) \times (6x)$

$$= \underline{6x\cos(3x^2+7)}$$

(e)  $y = \tan^4(2x+3)$  $\Rightarrow$  $\dfrac{dy}{dx} = 4\tan^3(2x+3) \times \sec^2(2x+3) \times 2$

$$= \underline{8\tan^3(2x+3)\sec^2(2x+3)}$$

Here we have implicitly used a slightly extended version of the chain rule $dy/dx = dy/du \times du/dv \times dv/dx$ where $v = 2x+3$, $u = \tan v$ and $y = u^4$; the result then follows from the substitution of the derivatives $dy/du = 4u^3$, $du/dv = \sec^2 v$ and $dv/dx = 2$.

(f)  $y = x\,e^{-3x^2}$  $\Rightarrow$  $\dfrac{dy}{dx} = x\,e^{-3x^2} \times (-6x) + e^{-3x^2}$

$$= \underline{(1-6x^2)\,e^{-3x^2}}$$

This question involves the use of the product rule, $(uv)' = uv' + u'v$, where $u = x$ and $v = e^{-3x^2}$; the evaluation of $v'$ itself entails the chain rule, with $w = -3x^2$ and $v = e^w$.

(g)  $y = x\ln(x^2+1)$  $\Rightarrow$  $\dfrac{dy}{dx} = x \times \dfrac{1}{(x^2+1)} \times 2x + \ln(x^2+1)$

$$= \frac{2x^2}{(x^2+1)} + \ln(x^2+1)$$

(h)  $y = \dfrac{\sin x}{x}$  $\Rightarrow$  $\dfrac{dy}{dx} = \dfrac{x\cos x - \sin x}{x^2}$

This is an example of the use of the quotient rule, $(u/v)' = (vu' - uv')/v^2$, with $u = \sin x$ and $v = x$.

**(3)** By exploiting the linearity property of a differential operator, and using a knowledge of the sum of an infinite GP, show that

$$\sum_{n=1}^{\infty} n\,x^{n-1} = \frac{\mathrm{d}}{\mathrm{d}x}\left(\frac{x}{1-x}\right)$$

where $|x| < 1$. Hence, evaluate the summation.

$$\sum_{n=1}^{\infty} n\,x^{n-1} = \sum_{n=1}^{\infty} \frac{\mathrm{d}}{\mathrm{d}x}(x^n) = \frac{\mathrm{d}}{\mathrm{d}x}\left(\sum_{n=1}^{\infty} x^n\right)$$

$$= \frac{\mathrm{d}}{\mathrm{d}x}\left(\frac{x}{1-x}\right) \quad \text{for } |x| < 1$$

$$= \frac{1-x+x}{(1-x)^2} = \frac{1}{(1-x)^2}$$

---

**(4)** Differentiate $y = \cos^{-1}(x/a)$, where $|x/a| < 1$; is the result valid for all values of $y$? What is the derivative of $\tan^{-1}(x/a)$?

If $y = \cos^{-1}(x/a)$, then $x = a\cos y$   (for $|x/a| < 1$)

$$\therefore \quad \frac{\mathrm{d}x}{\mathrm{d}y} = -a\sin y = -a\sqrt{1-\cos^2 y} = -a\sqrt{1-x^2/a^2} = -\sqrt{a^2-x^2}$$

$$\therefore \quad \frac{\mathrm{d}y}{\mathrm{d}x} = \frac{1}{\mathrm{d}x/\mathrm{d}y} = \frac{-1}{\sqrt{a^2-x^2}}$$

i.e. $\quad \dfrac{\mathrm{d}}{\mathrm{d}x}\left[\cos^{-1}(x/a)\right] = \dfrac{-1}{\sqrt{a^2-x^2}}$

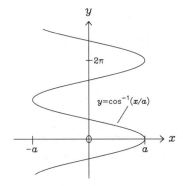

This formula is correct for $0 \leqslant y \leqslant \pi$, and for multiple additions of $2\pi$; in other words, for $2n\pi \leqslant y \leqslant (2n+1)\pi$ where $n$ is an integer. If $\pi < y < 2\pi$, for example, then the derivative of $\cos^{-1}(x/a)$ is $+1/\sqrt{a^2-x^2}$.

$$y = \tan^{-1}(x/a) \quad \Longleftrightarrow \quad x = a\tan y$$

$$\therefore \quad \frac{\mathrm{d}x}{\mathrm{d}y} = a\sec^2 y = a\left(1+\tan^2 y\right) = a\left(1+x^2/a^2\right) = \left(a^2+x^2\right)/a$$

$$\therefore \quad \frac{\mathrm{d}}{\mathrm{d}x}\left[\tan^{-1}(x/a)\right] = \frac{a}{a^2+x^2}$$

**(5)** Find $dy/dx$ when: (a) $x = t(t^2+2)$ and $y = t^2$; (b) $x^2 = y\sin(xy)$.

(a)  $x = t^3 + 2t \;\Rightarrow\; \dfrac{dx}{dt} = 3t^2 + 2$

$y = t^2 \;\Rightarrow\; \dfrac{dy}{dt} = 2t$

$\left.\vphantom{\begin{array}{c}a\\b\end{array}}\right\}$  $\therefore \dfrac{dy}{dx} = \dfrac{dy/dt}{dx/dt} = \dfrac{2t}{(3t^2+2)}$

(b)  $x^2 = y\sin(xy) \;\Rightarrow\; \dfrac{d}{dx}(x^2) = \dfrac{d}{dx}\big[y\sin(xy)\big]$

$\therefore\; 2x = y\dfrac{d}{dx}\big[\sin(xy)\big] + \sin(xy)\dfrac{d}{dx}(y)$

$= y\cos(xy)\dfrac{d}{dx}(xy) + \sin(xy)\dfrac{dy}{dx}$

$= y\cos(xy)\Big[x\dfrac{dy}{dx} + y\Big] + \sin(xy)\dfrac{dy}{dx}$

$\therefore\; 2x = y^2\cos(xy) + \dfrac{dy}{dx}\big[xy\cos(xy) + \sin(xy)\big]$

i.e.  $\dfrac{dy}{dx} = \dfrac{2x - y^2\cos(xy)}{xy\cos(xy) + \sin(xy)}$

---

**(6)** Find and classify the stationary points of the following functions of $x$ and $r$, where $\epsilon$, $\sigma$ and $a_0$ are constants.

(a)  $f(x) = \dfrac{x^5}{5} - \dfrac{x^4}{6} - x^3 \qquad \therefore\; f'(x) = x^4 - \dfrac{2x^3}{3} - 3x^2$

$\therefore\; f''(x) = 4x^3 - 2x^2 - 6x$

For stationary points, $f'(x) = 0 \;\Rightarrow\; \frac{1}{3}x^2(3x^2 - 2x - 9) = 0$

$\therefore\; x = 0 \quad \text{or} \quad x = \dfrac{2 \pm \sqrt{4 + 108}}{6} = \dfrac{1 \pm 2\sqrt{7}}{3}$

When $x = 0$, $f''(x) = 0 \qquad \therefore\; f''(x)$ test is inconclusive

As $x \to 0$, $f(x) \to -x^3 \qquad \therefore\; \underline{x = 0 \text{ is a point of inflexion}}$

When $x = \frac{1-2\sqrt{7}}{3}$, $f''(x) < 0 \qquad \therefore\; \underline{x = \frac{1-2\sqrt{7}}{3} \text{ is a maximum}}$

When $x = \frac{1+2\sqrt{7}}{3}$, $f''(x) > 0 \qquad \therefore\; \underline{x = \frac{1+2\sqrt{7}}{3} \text{ is a minimum}}$

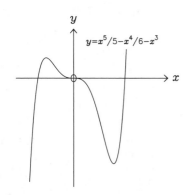

$y = x^5/5 - x^4/6 - x^3$

(b)    $f(x) = \dfrac{x}{1+x^2}$

$\therefore \ f'(x) = \dfrac{1+x^2-2x^2}{(1+x^2)^2} = \dfrac{1-x^2}{(1+x^2)^2}$

$\therefore \ f''(x) = \dfrac{-2x\,(1+x^2)^2 - 4x\,(1-x^2)(1+x^2)}{(1+x^2)^4} = \dfrac{-2x\,(3-x^2)}{(1+x^2)^3}$

For stationary points, $f'(x) = 0 \ \Rightarrow \ 1 - x^2 = 0$

$$\therefore \ x = \pm 1$$

$$f''(+1) = -\tfrac{1}{2} < 0 \qquad \therefore \ \underline{x = 1 \text{ is a maximum}}$$

$$f''(-1) = +\tfrac{1}{2} > 0 \qquad \therefore \ \underline{x = -1 \text{ is a minimum}}$$

(c)    $U(r) = 4\epsilon\left[\left(\dfrac{\sigma}{r}\right)^{12} - \left(\dfrac{\sigma}{r}\right)^6\right] \qquad$ for $\epsilon > 0$ and $r \geqslant 0$

$\therefore \ U'(r) = 4\epsilon\left[12\left(\dfrac{\sigma}{r}\right)^{11} - 6\left(\dfrac{\sigma}{r}\right)^5\right]\left(\dfrac{-\sigma}{r^2}\right) = -\dfrac{24\,\epsilon}{\sigma}\left[2\left(\dfrac{\sigma}{r}\right)^{13} - \left(\dfrac{\sigma}{r}\right)^7\right]$

$\therefore \ U''(r) = \dfrac{24\,\epsilon}{\sigma^2}\left[26\left(\dfrac{\sigma}{r}\right)^{14} - 7\left(\dfrac{\sigma}{r}\right)^8\right]$

For stationary points, $U'(r) = 0 \ \Rightarrow \ \left(\dfrac{\sigma}{r}\right)^7\left[1 - 2\left(\dfrac{\sigma}{r}\right)^6\right] = 0$

$$\therefore \ r \to \infty \quad \text{or} \quad r = 2^{1/6}\sigma$$

As $r \to \infty$, $U(r)$ decays to zero $\quad \therefore$ Not a proper stationary point.

$$U''(2^{1/6}\sigma) > 0 \qquad \therefore \ \underline{\text{Minimum at } r = 2^{1/6}\sigma}$$

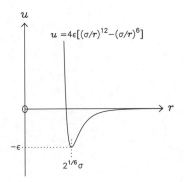

$u = 4\epsilon[(\sigma/r)^{12} - (\sigma/r)^6]$

The function $U(r)$ is known as the *Lennard-Jones 6-12 potential*, and is often used to approximate the potential energy between neutral molecules. While there are no interactions between the constituents of an *ideal gas*, there are in real ones: a short-range repulsion, which stops two molecules from occupying the same space; and a longer-range attraction (from a *van der Waals* type dipole-induced-dipole interaction) which weakens as the molecules move further apart. The equilibrium separation $r$ represents a balance between these opposing forces and is given by the location of the minimum in $U(r)$.

(d) $\quad p(r) = \dfrac{r^2}{8\,a_0^3}\left(2 - \dfrac{r}{a_0}\right)^2 e^{-r/a_0} \qquad$ for $\;a_0 > 0\;$ and $\;r \geqslant 0$

$$\therefore\; p'(r) = \dfrac{1}{8\,a_0^3}\left[2r\left(2 - \dfrac{r}{a_0}\right)^2 - \dfrac{2r^2}{a_0}\left(2 - \dfrac{r}{a_0}\right) - \dfrac{r^2}{a_0}\left(2 - \dfrac{r}{a_0}\right)^2\right]e^{-r/a_0}$$

$$= \dfrac{r}{8\,a_0^3}\left(2 - \dfrac{r}{a_0}\right)\left[\left(\dfrac{r}{a_0}\right)^2 - 6\dfrac{r}{a_0} + 4\right]e^{-r/a_0}$$

$$\therefore\; p''(r) = \dfrac{1}{8\,a_0^3}\left\{\left(2 - \dfrac{r}{a_0}\right)\left[\left(\dfrac{r}{a_0}\right)^2 - 6\dfrac{r}{a_0} + 4\right]\right.$$

$$- \dfrac{r}{a_0}\left[\left(\dfrac{r}{a_0}\right)^2 - 6\dfrac{r}{a_0} + 4\right]$$

$$+ \dfrac{r}{a_0}\left(2 - \dfrac{r}{a_0}\right)\left[2\dfrac{r}{a_0} - 6\right]$$

$$\left. - \dfrac{r}{a_0}\left(2 - \dfrac{r}{a_0}\right)\left[\left(\dfrac{r}{a_0}\right)^2 - 6\dfrac{r}{a_0} + 4\right]\right\}e^{-r/a_0}$$

$$= \dfrac{1}{8\,a_0^3}\left[\left(\dfrac{r}{a_0}\right)^4 - 12\left(\dfrac{r}{a_0}\right)^3 + 40\left(\dfrac{r}{a_0}\right)^2 - 40\dfrac{r}{a_0} + 8\right]e^{-r/a_0}$$

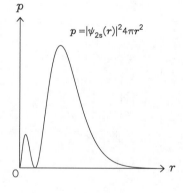

For stationary points, $\;p'(r) = 0$

$$\therefore\quad r = 0, \quad r = 2\,a_0, \quad \dfrac{r}{a_0} = \dfrac{6 \pm \sqrt{36 - 16}}{2} = 3 \pm \sqrt{5}\;\text{ or }\; r \to \infty$$

When $\;r = 0\;$ or $\;r \to \infty$, they are not proper stationary points

$$p''(2\,a_0) > 0 \qquad \therefore\; r = 2\,a_0 \text{ is a minimum}$$

$$p''\big[(3 - \sqrt{5})a_0\big] < 0 \qquad \therefore\; r = (3 - \sqrt{5})\,a_0 \text{ is a maximum}$$

$$p''\big[(3 + \sqrt{5})a_0\big] < 0 \qquad \therefore\; r = (3 + \sqrt{5})\,a_0 \text{ is a maximum}$$

The function $p(r)$ is the radial probability density function for an electron in the $2s$ state of a hydrogen atom; that is, the probability of finding the electron in a small interval between $r$ and $r + \delta r$ from the nucleus is given by $p(r)\,\delta r$ (where $p(r) = 4\pi r^2 |\psi_{2s}|^2$). As well as the main maximum at $r = (3 + \sqrt{5})\,a_0$, where the electron is most probably to be found, there is also a smaller maximum closer to the nucleus at $r = (3 - \sqrt{5})\,a_0$; this subsidiary maximum is indicative of the *penetration effect*, and gives an important insight into the behaviour of electrons in atomic shells.

# 5 Integration

## 5.1 Areas and integrals

In chapter 4, we were concerned with looking at the slope of a curve $y = f(x)$; that is, in studying the rate at which two quantities varied with respect to each other. Now let us move on to the topic of *integration* which deals with the 'area under a curve'; this is intimately linked with the important questions of the average and cumulative behaviour of $y$, over some range in $x$.

To set up a formal definition of an integral, consider the region bounded by the curve $y = f(x)$ and the straight lines $x = a$, $x = b$, and $y = 0$. The size of the enclosure can be estimated by approximating it as a series of narrow vertical strips and adding together the areas of these contiguous rectangular blocks. If the $x$-axis between $a$ and $b$ is divided into N equal intervals, then the width of each strip is given by $\delta x = (b-a)/\text{N}$; the corresponding heights of the thin blocks are equal to the values of the function $f(x)$ at their central positions. In other words, the area of the $j^{\text{th}}$ strip, which is at $x = x_j$ and of height $y = f(x_j)$, is $f(x_j)\,\delta x$; the index $j$ ranges from 1 to N, of course, with $x_1 = a + \delta x/2$ and $x_\text{N} = b - \delta x/2$. As N tends to infinity, $\delta x \to 0$ and the approximation to the area under the curve becomes ever more accurate. This limiting form of the summation procedure defines an integral

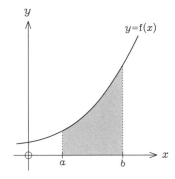

$$\int_a^b y\,\mathrm{d}x = \int_a^b f(x)\,\mathrm{d}x = \lim_{\text{N}\to\infty}\sum_{j=1}^{\text{N}} f(x_j)\,\delta x \qquad (5.1)$$

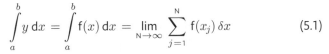

where the symbol $\int \mathrm{d}x$ is read as the 'integral, from $a$ to $b$, with respect to $x$'. The use of the term 'area' above needs some qualification, in that it can be negative; this is because the 'height' of a strip $f(x_j) < 0$ whenever the curve $y = f(x)$ lies below the $x$-axis (and even the 'width' $\delta x < 0$ if $b < a$).

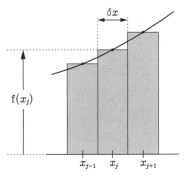

Before going on to see how integrals are evaluated, we should try to get a better feel for what they represent physically. As a simple example, suppose that a car moved in a straight line with a constant speed $v_0$ for a time $t_0$; then, it will have travelled a distance $v_0 t_0$. If the journey had progressed with a varying speed $v(t)$, as a function of time, how far would the car have gone? To work this out, the duration of the trip can be divided into a series of short intervals $\delta t$ over which the speed is roughly constant. The distance travelled is then given by the sum of all the small contributions $v(t)\,\delta t$; that is to say, the integral of $v(t)$, with respect to $t$, from $t=0$ to $t=t_0$

$$\text{Distance travelled} = \int_0^{t_0} v(t)\,dt \qquad (5.2)$$

Negative values of $v(t)$ can be understood as a backwards motion of the car, so that we should really talk about *velocity* rather than speed, and gives rise to a reduction in the (forward) length of the journey. The central rôle of an integral in the calculation of an average can also be appreciated, as the mean speed is the ratio of the overall distance travelled to the total time taken; in other words, the constant speed, $\langle v \rangle$ or $\bar{v}$, required for an equivalent trip is given by eqn (5.2) divided by $t_0$.

$$\bar{y} = \langle y \rangle = \frac{1}{b-a} \int_a^b y\,dx$$

## 5.2 Integrals and derivatives

Although an integral is defined as the limiting form of a summation, it is rarely ascertained from eqn (5.1). Rather, it's usually evaluated by noting that 'integration is the reverse of differentiation'. While this may not be obvious, it is easily confirmed for our earlier kinematic case: the distance travelled was seen to be the integral of the speed, and speed is the rate-of-change of distance (a derivative). To see this relationship more generally, consider the integral of $y=f(t)$ between the origin ($t=0$) and an arbitrary point $t=x$

$$\int_0^x y\,dt = G(x) \qquad (5.3)$$

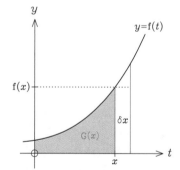

We have set this equal to a function $G(x)$ because the area under the curve will vary smoothly with the position of the upper bound. In particular, a small increment $\delta x$ will add a contribution of approximately $f(x)\,\delta x$

$$G(x+\delta x) \approx G(x) + f(x)\,\delta x$$

This expression becomes exact in the limit $\delta x \to 0$, where upon an algebraic rearrangement, and eqn (4.1), shows that

$$f(x) = \lim_{\delta x \to 0} \frac{G(x+\delta x) - G(x)}{\delta x} = \frac{dG}{dx} \qquad (5.4)$$

Hence, $G(x)$ is the function whose differential coefficient is equal to $f(x)$. Since the derivative of $\sin x$ is $\cos x$, for example, the integral of $\cos x$ is $\sin x$. Having obtained $G(x)$ in this manner, a simple consideration of the areas that $G(a)$ and $G(b)$ represent allows us to write the integral of eqn (5.1) as

$$\int_a^b f(x)\,dx = G(b) - G(a) = \left[G(x)\right]_a^b \qquad (5.5)$$

where the square-bracket notation on the far right is a standard abbreviation for the difference of terms in the middle.

The calculation of most integrals hinges on eqns (5.4) and (5.5). That is to say: (i) thinking about differentiation backwards to obtain $G(x)$; (ii) the substitution of the limits $a$ and $b$. While the second step is straightforward, the former often involves some preliminary manipulations; we will come to these shortly. An important general point, however, is that eqn (5.4) only determines $G(x)$ to within an arbitrary additive constant $K$ (say); this is because $G(x)$ and $G(x)+K$ have the same differential coefficient since $dK/dx = 0$. This degree of uncertainty is of no consequence for eqn (5.5), as $K$ cancels out on taking the difference, but it does matter if the limits of the integration are not specified. The latter case is called an *indefinite* integral, in contrast to the *definite* ones of eqns (5.1) and (5.5), and requires an additional constraint, or *boundary condition*, such as the value of $G(0)$, to pin down the 'constant of integration'.

$$\int_1^2 x \, dx = \left[\frac{x^2}{2}\right]_1^2 = \frac{4}{2} - \frac{1}{2} = \frac{3}{2}$$

$$G(x) = \int \sin x \, dx = K - \cos x$$

If $G(0) = 0$, then $K = 1$

## 5.3 Some basic properties of integrals

Just as for the case of derivatives in section 4.2, one of the most fundamental properties of integrals is their linearity. That is to say, they satisfy the rules

$$\int A \, f(x) \, dx = A \int f(x) \, dx$$

and   $$\int \left[ f(x) + g(x) \right] dx = \int f(x) \, dx + \int g(x) \, dx \qquad (5.6)$$

where $A$ is a constant. This is useful because it allows us to work out the integral of a sum, or difference, of terms (e.g. a polynomial) by knowing the integrals of its constituent parts.

While eqn (5.6) holds for all integrals, there are several general results that apply specifically to definite ones. The first is that an interchange of the order of the limits reverses the sign of the integral, and follows from eqn (5.5)

$$\int_a^b f(x) \, dx = -\int_b^a f(x) \, dx \qquad (5.7)$$

Another formula emanating from eqn (5.5) concerns the decomposition of an integral into the sum of two, or more, sharing suitable common limits

$$\int_a^c f(x) \, dx = \int_a^b f(x) \, dx + \int_b^c f(x) \, dx \qquad (5.8)$$

This can be understood directly, at least for the case $a < b < c$, by thinking about the relevant areas under the curve $y = f(x)$. Finally, the formal inverse relationship between integration and differentiation can be exploited to yield

$$\frac{\mathrm{d}}{\mathrm{d}t}\left(\int_{a(t)}^{b(t)} \mathrm{f}(x)\,\mathrm{d}x\right) = \mathrm{f}(b)\frac{\mathrm{d}b}{\mathrm{d}t} - \mathrm{f}(a)\frac{\mathrm{d}a}{\mathrm{d}t} \qquad (5.9)$$

where we have implicitly used eqns (5.4) and (5.5), and the chain rule of eqn (4.12). Usually $a$ and $b$ are constants, and so the definite integral is simply a number (not a function of $x$, which is then a 'dummy variable') whose derivative is equal to zero. If the limits do depend on the value of a parameter, say $t$, then we would generally expect the area under the curve to vary with respect to it; although the related rate-of-change could be calculated by first evaluating the integral and then differentiating the result with respect to $t$, eqn (5.9) provides a short cut that avoids the need for explicitly carrying out the initial step and simplifies the latter.

If $G(t) = \int_{t}^{t^2} \sin x\,\mathrm{d}x$

$= \cos t - \cos t^2$

$\dfrac{\mathrm{d}G}{\mathrm{d}t} = 2t\sin t^2 - \sin t$

## 5.4 Inspection and substitution

As noted earlier, most integrals are evaluated by remembering that integration is the reverse of differentiation. Thus the backwards application of the results in chapter 4 gives rise to many 'standard integrals'. For example, eqns (4.7) and (4.8) yield

$$\int \mathrm{e}^x\,\mathrm{d}x = \mathrm{e}^x + K \quad \text{and} \quad \int \frac{1}{x}\,\mathrm{d}x = \ln x + K \qquad (5.10)$$

where $K$ is the constant of integration, and $x$ is assumed to be positive in the logarithmic case. The key step in the calculation of $\int \mathrm{f}(x)\,\mathrm{d}x$ is the recognition of the *integrand*, $\mathrm{f}(x)$, as the differential coefficient of a known function; this requires a spotting ability which is enhanced through experience, but initiated by acquiring a good familiarity with the derivatives of the previous chapter. Whenever possible, an integral should be checked by making sure that its derivative returns the integrand; this test eliminates many elementary errors.

$\int x^M\,\mathrm{d}x = \dfrac{x^{M+1}}{M+1} + K$

(for $M \neq -1$)

When $\int \mathrm{f}(x)\,\mathrm{d}x$ does not appear to be of a standard form, it can often be transformed into one through a suitable 'change of variables'. That is to say, the *substitution* of $u = \mathrm{g}(x)$ may lead to an integral of a function of $u$ which is recognized more easily. For example, the integral of $x\sqrt{4-x}$, with respect to $x$, becomes straightforward on putting $u = 4 - x$

$$\int x\sqrt{4-x}\,\mathrm{d}x = -\int (4-u)\sqrt{u}\,\mathrm{d}u = \int (u^{3/2} - 4u^{1/2})\,\mathrm{d}u$$

where the last step makes use of eqns (1.3) and (1.4). While the substitution of $x = 4 - u$ is obvious, the replacement of $\mathrm{d}x$ requires a bit of thought; it follows from the derivative $\mathrm{d}x/\mathrm{d}u = -1$, so that $\mathrm{d}x = -\mathrm{d}u$. If this splitting up of $\mathrm{d}x$ and $\mathrm{d}u$ seems strange, then the result can be viewed as the limiting form of the relationship $\delta x \approx \mathrm{d}x/\mathrm{d}u \times \delta u \approx -\delta u$. The linear property of eqn (5.6), and a knowledge of the integral of $u^M$, then gives

$$\int x\sqrt{4-x}\ \mathrm{d}x \ = \ \tfrac{2}{5}\,u^{5/2} - \tfrac{8}{3}\,u^{3/2} + K$$

where the right-hand side can be expressed in terms of $x$ by putting $u = 4 - x$. This 'back-substitution' is not needed for a definite integral if the limits $x = a$ and $x = b$ are applied in terms of their equivalent values in $u$; that is, $u = 4 - a$ and $u = 4 - b$.

There are no general rules for working out which substitution to use, or deciding whether one will be helpful at all, without actually trying it. There are several cases that are met frequently, however, and it is worth mentioning these explicitly. Firstly, consider the integral of $\sin^M x \cos^N x$ where M and N are integers. If M is odd, then putting $u = \cos x$ transforms the trigonometric integrand in $x$ to a simple polynomial one in $u$; remembering, of course, that $\mathrm{d}u = -\sin x\,\mathrm{d}x$ and even powers of $\sin x$ can be factored in terms of $1 - u^2$ from eqn (3.8). Similarly, putting $u = \sin x$ is helpful if N is odd. If both M and N are even, then the integration usually entails a lot more effort. In particular, it involves the repeated use of the double-angle formulae of section 3.4, especially eqn (3.15), to transform the integrand into a familiar form; as an example, try integrating $\sin^4 x$ with the aid of exercise 5 in section 3.8.

In integrals containing the term $a^2 - x^2$, where $a$ is a constant, it is often helpful to put $x = a\sin\theta$ or $x = a\cos\theta$; with factors of $a^2 + x^2$ and $x^2 - a^2$, the substitution $x = a\tan\theta$ and $x = a\sec\theta$ respectively is sometimes useful. As a last resort for integrals of trigonometric functions, it can be worth trying $t = \tan(x/2)$.

Finally, we should mention the case where the integrand takes the form of some function of $u = g(x)$ times $\mathrm{d}u/\mathrm{d}x$. The presence of the derivative simplifies the integral greatly since, according to the chain rule of section 4.5, we have

$$\int f(u)\,\frac{\mathrm{d}u}{\mathrm{d}x}\ \mathrm{d}x \ = \ \int f(u)\ \mathrm{d}u \qquad (5.11)$$

If this situation can be spotted then the explicit substitution $u = g(x)$ is not necessary and, with experience, the integral can often be evaluated directly in our head. For example, the integral of $x\,e^{-x^2}$ is equal to minus a half of $e^{-x^2}$ because, to within a factor of $-1/2$, the integrand is the product of 'e to the power of something and the derivative of the something'. We should also note that the prefactor $x$ is crucial in this instance as the integral of $e^{-x^2}$ itself is not at all straightforward!

$$\int_{0}^{\pi/2} \sin^4 x\,\cos^3 x\ \mathrm{d}x$$

$$= \int_{0}^{1} u^4(1-u^2)\ \mathrm{d}u$$

$$(\text{where }u = \sin x)$$

If $t = \tan\left(\frac{x}{2}\right) \quad \mathrm{d}x = \frac{2\,\mathrm{d}t}{1+t^2}$

$\sin x = \frac{2t}{1+t^2} \quad \cos x = \frac{1-t^2}{1+t^2}$

$$\int \frac{2x}{x^2-1}\ \mathrm{d}x \ = \ \ln(x^2-1) + K$$

## 5.5 Partial fractions

In section 1.7, we discussed the topic of partial fractions whereby an algebraic ratio is decomposed into the sum of several constituents parts. While this procedure seemed like a backwards step, it can be very useful for evaluating suitable types of integrals. As a simple illustration, consider the case

$$\int \frac{\mathrm{d}x}{x\,(x+1)} = \int \frac{\mathrm{d}x}{x} - \int \frac{\mathrm{d}x}{x+1}$$

where we have used a partial fraction analysis of $1/[x\,(x+1)]$, and the linear property of eqn (5.6), to obtain the difference of the two terms on the right-hand side; the latter integrals are easily recognized as being given by the logarithms of their respective denominators.

## 5.6 Integration by parts

The integration of eqn (4.9), which gives the rule for differentiating a product, leads to the general relationship

$$\int u \frac{\mathrm{d}v}{\mathrm{d}x}\,\mathrm{d}x = u\,v - \int v \frac{\mathrm{d}u}{\mathrm{d}x}\,\mathrm{d}x \qquad (5.12)$$

where u and v are arbitrary functions of $x$. This slightly odd-looking equation refers to the situation where the integrand consists of the product of two terms of which one, labelled as u, can be differentiated and the other, denoted by dv/d$x$, easily integrated. The use of eqn (5.12) is called *integration by parts*, with the idea being that the integral on the right-hand side might be more 'doable' than the original one on the left.

As an example, suppose that we wanted to evaluate $\int x \ln x\,\mathrm{d}x$. Then, since the integral of $x$ is $x^2/2$ and the derivative of $\ln x$ is $1/x$, eqn (5.12) gives

$$\int x \ln x\,\mathrm{d}x = \frac{x^2}{2} \ln x - \int \frac{x}{2}\,\mathrm{d}x$$

$$\int \ln x\,\mathrm{d}x = x \ln x - x + K$$

where the integral on the far right is equal to $x^2/4 + K$. On a related note, and somewhat surprisingly, the integral of $\ln x$ can be ascertained in this manner by expressing the integrand as $1 \times \ln x$.

## 5.7 Reduction formulae

Consider the following integral $I_n$, where the subscript $n$ corresponds to the power of $x$ in the integrand

$$I_n = \int_0^\infty x^n\,\mathrm{e}^{-x}\,\mathrm{d}x \qquad (5.13)$$

$$\int_0^\infty \mathrm{e}^{-x}\,\mathrm{d}x = \left[-\mathrm{e}^{-x}\right]_0^\infty = 1$$

If $n=0$, so that $x^n=1$, then it is simple to show that $I_0=1$. If $n$ is a large positive integer, then $I_n$ can be evaluated by relating it to $I_0$ through the repeated use of integration by parts. To see this, let us carry out this procedure on $I_n$ just once

$$I_n = \left[-x^n\,\mathrm{e}^{-x}\right]_0^\infty + n\int_0^\infty x^{n-1}\,\mathrm{e}^{-x}\,\mathrm{d}x$$

where we have integrated $e^{-x}$ and differentiated $x^n$. The first term on the right is equal to zero, because $x^n e^{-x}$ is nought at both the upper and lower limit; by the definition of eqn (5.13), the integral is $I_{n-1}$. Hence, we have

$$I_n = n I_{n-1} \tag{5.14}$$

If $n = 7$, for example, then $I_7 = 7 I_6$. Using eqn (5.14) several times over, we obtain $I_6 = 6 I_5$, $I_5 = 5 I_4$, and so on until we reach $I_1 = I_0 = 1$. Hence, on combining all these elements together, we see that $I_n$ is equal to $n$-factorial. This integral definition of $n!$ is called a *Gamma function*, $\Gamma(n+1)$, and confirms our assertion in section 1.5 that $0! = 1$.

$$\int_0^\infty x^n\, e^{-x}\, dx = n!$$

The main purpose of the calculation above was to illustrate the derivation and use of a *reduction formula*, such as eqn (5.14); these relate integrals of 'order $n$', generally denoted by $I_n$, to corresponding ones of lower 'powers'.

## 5.8 Symmetry, tables, and numerical integration

In concluding this chapter, we should reiterate that the successful evaluation of an integral hinges on our ability to spot an integrand as the differential coefficient of a known function. Procedures like substitution and integration by parts are often helpful in that they can transform an unfamiliar case into one that is easily recognized; several of these manipulations may be needed in any given problem, and sometimes more than once.

In real life, unlike exams, the best way of proceeding is usually to look up the result in a 'handbook of integrals'. Even with the aid of these reference works, there are many situations when an integral cannot be done; that is to say, it does not have a simple algebraic form. A definite integral can always be computed numerically, of course, by calculating the area under the curve by using the summation of eqn (5.1); commonly met cases, such as $\int e^{-x^2} dx$ between zero and a variable upper limit (which is called an *error function*), are frequently found tabulated in books.

Finally, we should mention that some integrals can be set to zero based solely on the symmetry of their integrand. This is especially true of *antisymmetric* or *odd* functions, such as $f(x) = x$ and $f(x) = \sin x$, when they are integrated over a symmetric range around the origin

$$\text{If } f(-x) = -f(x), \quad \int_{-a}^{a} f(x)\, dx = 0 \tag{5.15}$$

This property can be proved algebraically, with eqns (5.7) and (5.8), or reasoned graphically: as the curve $y = f(x)$ on one side of the line $x = 0$ is an inverted mirror image of that on the other, the integral is zero because the areas in the two halves have the same magnitude but opposite signs. The counterpart of eqn (5.15) for *symmetric* or *even* functions, such as $f(x) = x^2$ and $f(x) = \cos x$, where $f(-x) = f(x)$, is that the integral from $-a$ to $a$ is equal to twice that from 0 to $a$. Since most functions are neither odd nor even, but are of a mixed symmetry, the above discussion does not apply to them.

## 5.9 Worked examples

> **(1)** Integrate the following functions with respect to $x$:
> (a) $x + \sqrt{x} - 1/x$,   (b) $\sqrt{x}\,(x - 1/x)$,   (c) $2^x$,   (d) $e^{2x}$,
> (e) $1/(2x - 1)$,   (f) $\sin 2x + \cos 3x$,   (g) $\tan x$,   (h) $\sin^2 x$,
> (i) $x/(1 + x^2)$,   (j) $1/(1 + x^2)$.

(a) $\displaystyle \int \left( x + \sqrt{x} - \frac{1}{x} \right) dx = \int \left( x + x^{1/2} - x^{-1} \right) dx$

$$= \tfrac{1}{2}x^2 + \tfrac{2}{3}x^{3/2} - \ln x + K$$

$$= \tfrac{1}{2}x^2 + \tfrac{2}{3}x\sqrt{x} - \ln x + K$$

(b) $\displaystyle \int \sqrt{x}\left( x - \frac{1}{x} \right) dx = \int \left( x^{3/2} - x^{-1/2} \right) dx$

$$= \tfrac{2}{5}x^{5/2} - 2\,x^{1/2} + K$$

$$= 2\sqrt{x}\left( \tfrac{1}{5}x^2 - 1 \right) + K$$

(c) $\displaystyle \int 2^x \, dx = \frac{2^x}{\ln 2} + K$

(d) $\displaystyle \int e^{2x} \, dx = \tfrac{1}{2}e^{2x} + K$

(e) $\displaystyle \int \frac{dx}{2x - 1} = \tfrac{1}{2}\ln(2x - 1) + K$

If the integral above is not obvious, then the implicit manipulation can be made explicit with the substitution $u = 2x - 1$ (giving $du = 2\,dx$):

$$\int \frac{dx}{2x - 1} = \int \frac{du/2}{u} = \tfrac{1}{2}\int \frac{du}{u} = \tfrac{1}{2}\ln u + K$$

(f) $\displaystyle \int (\sin 2x + \cos 3x)\, dx = -\tfrac{1}{2}\cos 2x + \tfrac{1}{3}\sin 3x + K$

(g) $\displaystyle \int \tan x \, dx = \int \frac{\sin x}{\cos x}\, dx = -\ln(\cos x) + K = \ln(\sec x) + K$

Having expressed $\tan x$ as $\sin x/\cos x$, the integral becomes easy because the numerator is equal to the derivative of the denominator (give or take a minus sign). If the significance of the latter is not apparent in terms of the chain rule of differentiation, then the result can be obtained with the substitution $u = \cos x$ (whence $du = -\sin x\, dx$):

$$\int \frac{\sin x}{\cos x}\, dx = -\int \frac{du}{u} = -\ln u + K$$

(h)  $\int \sin^2 x\, dx = \tfrac{1}{2}\int (1 - \cos 2x)\, dx = \tfrac{1}{2}\left(x - \tfrac{1}{2}\sin 2x\right) + K$

(i)  $\int \frac{x}{1+x^2}\, dx = \tfrac{1}{2}\ln(1+x^2) + K$

This is again an easy integral because the numerator is equal to the derivative of the denominator, to within a constant multiplicative factor; the formal substitution, if required, would be $u = 1 + x^2$ (and $du = 2x\, dx$).

(j)  $\int \frac{dx}{1+x^2} = \tan^{-1}x + K$

Although we have simply stated the answer, because it is a 'standard integral', it can be ascertained systematically by substituting $x = \tan\theta$. Then, using the derivative $dx = \sec^2\theta\, d\theta = (1 + \tan^2\theta)\, d\theta = (1 + x^2)\, d\theta$, we are led to the result above:

$$\int \frac{dx}{1+x^2} = \int d\theta = \theta + K = \tan^{-1}x + K$$

---

(2) The ratio of peak areas in a $^1H$ nuclear magnetic resonance spectrum corresponds to the relative number of hydrogen atoms in each environment. Given that the peaks are Lorentzian in shape

$$f(x) = \frac{A}{1 + [\,2(x-x_0)/W\,]^2}$$

where $A$, $x_0$, and $W$ are their amplitude, location, and half-height width respectively, calculate an expression for the area under the peaks.

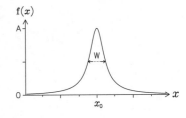

Putting  $u = \tfrac{2}{W}(x - x_0)$,  so that  $du = \tfrac{2}{W}\, dx$

$$\text{Area} = \int_{-\infty}^{\infty} f(x)\, dx = \frac{AW}{2}\int_{-\infty}^{\infty}\frac{du}{1+u^2} = \frac{AW}{2}\left[\tan^{-1}u\right]_{-\infty}^{\infty} = \frac{\pi AW}{2}$$

$$\tfrac{\pi}{2} + n\pi - \left(-\tfrac{\pi}{2} + n\pi\right) = \pi$$

> **(3)** Evaluate the following definite integrals:
>
> (a) $\displaystyle\int_0^{\pi/2} \sin^4 x \, \cos x \, dx$,    (b) $\displaystyle\int_0^{\pi/2} \sin^4 x \, dx$,    (c) $\displaystyle\int_0^4 \frac{x+3}{\sqrt{2x+1}} \, dx$.

(a)  $\displaystyle\int_0^{\pi/2} \sin^4 x \, \cos x \, dx = \left[\tfrac{1}{5}\sin^5 x\right]_0^{\pi/2} = \tfrac{1}{5}\left[\sin^5\left(\tfrac{\pi}{2}\right) - \sin^5(0)\right] = \tfrac{1}{5}$

The integral of $\sin^4 x \, \cos x$ is easy to write down because of the presence of a $\cos x$ with the function of $\sin x$; the manipulation carried out mentally above can be made explicit with the substitution $u = \sin x$ and $du = \cos x \, dx$, leading to $\int \sin^4 x \, \cos x \, dx = \int u^4 \, du$. If the derivative of $\sin x$ had not been present in the integrand, then the sum would have entailed a lot more effort; in particular, the (repeated) use of the double-angle formula $\cos 2x = 1 - 2\sin^2 x$.

(b)  From section 3.8 (example 5),  $8\sin^4\theta = \cos 4\theta - 4\cos 2\theta + 3$.

$$\therefore \quad \int_0^{\pi/2} \sin^4 x \, dx = \tfrac{1}{8}\int_0^{\pi/2} (\cos 4x - 4\cos 2x + 3) \, dx$$

$$= \tfrac{1}{8}\left[\tfrac{1}{4}\sin 4x - 2\sin 2x + 3x\right]_0^{\pi/2}$$

$$= \tfrac{1}{8}\left[\tfrac{1}{4}(\sin 2\pi - \sin 0) - 2(\sin \pi - \sin 0) + 3\left(\tfrac{\pi}{2} - 0\right)\right]$$

$$= \tfrac{3\pi}{16}$$

(c)  Let  $\displaystyle I = \int_0^4 \frac{x+3}{\sqrt{2x+1}} \, dx$

Put  $u^2 = 2x + 1$    $\therefore$   $2u \, du = 2 \, dx$

$$\therefore \quad I = \int_1^3 \frac{\tfrac{1}{2}(u^2 - 1) + 3}{u} \, u \, du = \int_1^3 \left(\tfrac{1}{2}u^2 + \tfrac{5}{2}\right) du$$

$$= \left[\tfrac{1}{6}u^3 + \tfrac{5}{2}u\right]_1^3$$

$$= \tfrac{1}{6}(27 - 1) + \tfrac{5}{2}(3 - 1)$$

$$= \tfrac{13}{3} + 5 = 9\tfrac{1}{3}$$

An alternative way of proceeding with this calculation is to use 'integration by parts', where $(2x+1)^{-1/2}$ is integrated and $x+3$ is differentiated:

$$\int_0^4 \frac{x+3}{\sqrt{2x+1}} \, dx = \left[(x+3)\sqrt{2x+1}\right]_0^4 - \int_0^4 \sqrt{2x+1} \, dx$$

$$= 21 - 3 - \left[\tfrac{1}{3}(2x+1)^{3/2}\right]_0^4$$

$$= 18 - \tfrac{1}{3}\left[9^{3/2} - 1\right]$$

$$= 18 - \tfrac{26}{3} = \underline{9\tfrac{1}{3}}$$

---

**(4)** Integrate the three partial fraction expressions in example 7 of section 1.8, with respect to $x$.

(a)  $\displaystyle\int \frac{dx}{x^2 - 5x + 6} = \int \frac{dx}{(x-3)} - \int \frac{dx}{(x-2)}$

$$= \ln(x-3) - \ln(x-2) + K = \underline{\ln\left(\tfrac{x-3}{x-2}\right) + K}$$

(b)  $\displaystyle\int \frac{x^2 - 5x + 1}{(x-1)^2(2x-3)} \, dx = 3\int \frac{dx}{(x-1)^2} + 9\int \frac{dx}{x-1} - 17\int \frac{dx}{2x-3}$

$$= \underline{K - \tfrac{3}{x-1} + 9\ln(x-1) - \tfrac{17}{2}\ln(2x-3)}$$

(c)  Let  $I = \displaystyle\int \frac{11x+1}{(x-1)(x^2 - 3x - 2)} \, dx$

$$= \int \frac{3x+5}{x^2 - 3x - 2} \, dx - 3\int \frac{dx}{x-1}$$

$$= \tfrac{3}{2}\int \frac{2x-3}{x^2 - 3x - 2} \, dx + \tfrac{19}{2}\int \frac{dx}{x^2 - 3x - 2} - 3\ln(x-1)$$

$$= \tfrac{3}{2}\ln(x^2 - 3x - 2) + \tfrac{19}{2}\int \frac{dx}{(x-\alpha)^2 - \gamma^2} - 3\ln(x-1)$$

where $\alpha = 3/2$ and $\gamma = \sqrt{17}/2$. The two logarithmic terms can be combined using eqns (1.7) and (1.8), and the denominator of the remaining integral factorised to enable a partial fraction decomposition.

$$\int \frac{dx}{(x-\alpha-\gamma)(x-\alpha+\gamma)} = \frac{1}{2\gamma}\left[\int \frac{dx}{x-\alpha-\gamma} - \int \frac{dx}{x-\alpha+\gamma}\right]$$

$$= \frac{1}{2\gamma}\ln\left(\frac{x-\alpha-\gamma}{x-\alpha+\gamma}\right) + K$$

$$\therefore \quad I = \frac{3}{2}\ln\left[\frac{x^2-3x-2}{(x-1)^2}\right] + \frac{19}{2\sqrt{17}}\ln\left[\frac{2x-3-\sqrt{17}}{2x-3+\sqrt{17}}\right] + K$$

Like many cases in real life, the evaluation of this integral has involved several manipulations. Starting with a partial fraction decomposition, transcribed from 7 (c) of section 1.8, and followed by the rewriting of the numerator $3x + 5$ as $\frac{3}{2}(2x - 3) + \frac{19}{2}$, so that the first part becomes a recognizable integral, to the completion of the squares in the denominator $x^2 - 3x - 2$ to eventually reach the final result.

---

**(5)** By integrating by parts, show that

$$\int \sin x\, e^{-x}\, dx = K - \frac{1}{2}(\sin x + \cos x)\, e^{-x}$$

---

Let $\quad I = \int \sin x\, e^{-x}\, dx$

$$= -\sin x\, e^{-x} + \int \cos x\, e^{-x}\, dx$$

$$= -\sin x\, e^{-x} - \cos x\, e^{-x} - \int \sin x\, e^{-x}\, dx$$

$$\therefore \quad I = -(\sin x + \cos x)\, e^{-x} - I$$

i.e. $\quad \int \sin x\, e^{-x}\, dx = K - \frac{1}{2}(\sin x + \cos x)\, e^{-x}$

---

**(6)** The probability that a system in equilibrium at temperature T is in a state with energy $E$ is given by the Boltzmann factor $e^{-E/kT}$, where k is the Boltzmann constant. Calculate the expected energy

$$\langle E \rangle = \int_0^\infty E\, g(E)\, e^{-E/kT}\, dE \bigg/ \int_0^\infty g(E)\, e^{-E/kT}\, dE$$

where the *density-of-states*, $g(E)$, is proportional to $E^\theta$ with $\theta \geqslant 0$.

Let $\quad I = \int\limits_0^\infty E\, E^\theta\, e^{-E/kT}\, dE$

$$= \underbrace{\left[ -kT\, E^{1+\theta}\, e^{-E/kT} \right]_0^\infty}_{=\,0} + kT\,(1+\theta) \int\limits_0^\infty E^\theta\, e^{-E/kT}\, dE$$

$$\lim_{x\to\infty} x^\alpha\, e^{-\beta x} = 0$$

(for $\beta > 0$)

But $\quad I = \langle E \rangle \int\limits_0^\infty E^\theta\, e^{-E/kT}\, dE \qquad \therefore \quad \underline{\langle E \rangle = kT\,(1+\theta)}$

---

**(7)** If

$$I_n = \int\limits_0^{\pi/2} \sin^n x\; dx$$

show that $n\, I_n = (n-1)\, I_{n-2}$, for $n \geqslant 1$. Hence, by relating odd orders of $I_n$ to $I_1$ and even ones to $I_0$, evaluate $I_5$ and $I_8$.

---

$$I_n = \int\limits_0^{\pi/2} \sin^{n-1} x\, \sin x\; dx \qquad (\text{for } n \geqslant 1)$$

$$= \left[ -\sin^{n-1} x\, \cos x \right]_0^{\pi/2} + (n-1) \int\limits_0^{\pi/2} \sin^{n-2} x\, \cos^2 x\; dx$$

$$= 0 + (n-1) \int\limits_0^{\pi/2} \sin^{n-2} x\, (1-\sin^2 x)\; dx$$

$$= (n-1) \left[ \int\limits_0^{\pi/2} \sin^{n-2} x\; dx - \int\limits_0^{\pi/2} \sin^n x\; dx \right]$$

$$\therefore \quad I_n = (n-1) \left[ I_{n-2} - I_n \right]$$

i.e. $\quad \underline{n\, I_n = (n-1)\, I_{n-2}} \quad (\text{for } n \geqslant 1)$

$$\therefore \quad \underline{I_5} = \tfrac{4}{5} I_3 = \tfrac{4}{5} \times \tfrac{2}{3} I_1 = \tfrac{8}{15}$$

$$I_n = \left( \tfrac{n-1}{n} \right) I_{n-2}$$

$$\therefore \quad \underline{I_8} = \tfrac{7}{8} I_6 = \tfrac{7}{8} \times \tfrac{5}{6} I_4 = \tfrac{7}{8} \times \tfrac{5}{6} \times \tfrac{3}{4} \times \tfrac{1}{2} I_0 = \tfrac{35\pi}{256}$$

$$\left[ I_1 = \int\limits_0^{\pi/2} \sin x\; dx = \left[ -\cos x \right]_0^{\pi/2} = 1, \quad I_0 = \int\limits_0^{\pi/2} dx = \left[ x \right]_0^{\pi/2} = \tfrac{\pi}{2} \right]$$

**(8)** Prove algebraically that $\int_{-a}^{a} f(x)\,dx = 0$ if $f(-x) = -f(x)$, and derive the corresponding expression for the symmetric function.

$$\int_{-a}^{a} f(x)\,dx = \int_{-a}^{0} f(x)\,dx + \int_{0}^{a} f(x)\,dx$$

$$= -\int_{a}^{0} f(-u)\,du + \int_{0}^{a} f(x)\,dx \qquad \text{(where } x = -u\text{)}$$

$$= \int_{a}^{0} f(u)\,du + \int_{0}^{a} f(x)\,dx \qquad \big[f(-u) = -f(u)\big]$$

$$= -\int_{0}^{a} f(u)\,du + \int_{0}^{a} f(x)\,dx = \underline{0}$$

In this example, we began by splitting up the integral from $-a$ to $+a$ into two parts (for positive $x$ and negative $x$); then we made the substitution $u = -x$ (so that $du = -dx$), used the fact that the integrand was antisymmetric, and interchanged the order of the limits. This eventually led to the difference of two identical integrals, even though one was written in terms of $x$ and the other in $u$. The reason that the $x$ and $u$ discrepancy does not matter is because they are 'dummy variables'; that is to say, the result of the definite integral is purely a number (dependent on $a$) and not on a function of $x$ or $u$.

If the integrand is symmetric, so that $f(-x) = f(x)$, then a retracing of the steps above leads to

$$\int_{-a}^{a} f(x)\,dx = \int_{0}^{a} f(u)\,du + \int_{0}^{a} f(x)\,dx = 2\int_{0}^{a} f(x)\,dx$$

Incidentally, any function can be rewritten as the sum of a symmetric (or even) contribution and an antisymmetric (or odd) one:

$$f(x) = \underbrace{\tfrac{1}{2}\big[f(x) + f(-x)\big]}_{f_{even}(x)} + \underbrace{\tfrac{1}{2}\big[f(x) - f(-x)\big]}_{f_{odd}(x)}$$

# 6 Taylor series

## 6.1 Approximating functions

When dealing with a complicated function, it can be useful to approximate it with one of a simpler form. While the latter may not represent a complete and accurate description of the situation at hand, it frequently provides the only means of making analytical progress. There are many approximations that could be used, of course, but it is the one that captures the salient features that is most helpful. In this chapter we will focus on the *Taylor series*, which is appropriate when our principal interest lies in the behaviour of a function in the neighbourhood of a particular point.

Consider the curve $y = f(x)$. The crudest approximation to this function is a horizontal line $y = a_0$, where $a_0$ is a constant; if $a_0 = f(x_0)$, then it will even be correct at $x = x_0$. A better approximation would be $y = a_0 + a_1(x - x_0)$, or a sloping line, where the coefficient $a_1$ allows for a non-zero gradient; if $a_1 = 0$, then we return to the earlier case. Continuing along this path, we could add a quadratic (or curvature) term $a_2(x - x_0)^2$, a cubic contribution $a_3(x - x_0)^3$, and so on, to gain further improvements. Thus, a function $f(x)$ can be approximated about the point $x_0$ by using a polynomial expansion

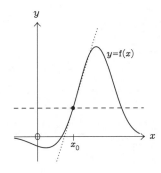

$$f(x) \approx a_0 + a_1(x - x_0) + a_2(x - x_0)^2 + a_3(x - x_0)^3 + \cdots \qquad (6.1)$$

This is, in fact, the essence of a Taylor series. Before dealing with the details of the coefficients, we should point out that the advantage of eqn (6.1) is that its right-hand side is usually easier to calculate, differentiate, integrate, and generally manipulate, than the expression on the left.

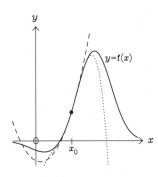

## 6.2 Derivation of the Taylor series

As we shall see shortly, the approximation of eqn (6.1) becomes an equality when enough terms are added and the magnitude of the difference $x - x_0$ is not too large. The substitution $x = x_0$ then gives $f(x_0) = a_0$, because everything involving $x - x_0$ on the right-hand side is equal to zero. If we differentiate eqn (6.1) with respect to $x$ and then put $x = x_0$, we obtain $f'(x_0) = a_1$. Repeating this procedure many times over, we find that $f''(x_0) = 2a_2$, $f'''(x_0) = 6a_3$, and, in

$f^n(x_0) = n!\, a_n$

general, that the value of the $n^{\text{th}}$ derivative of $f(x)$ at $x = x_0$, $f^n(x_0)$, is equal to $n!$ times $a_n$. Hence, the Taylor series can be written concisely as

$$f(x) = \sum_{n=0}^{\infty} \frac{f^n(x_0)}{n!} (x - x_0)^n \tag{6.2}$$

An alternative way of expressing this sum, which is sometimes more useful, is obtained by making the substitution $\epsilon = x - x_0$; explicitly, this gives

$$f(x_0 + \epsilon) = f(x_0) + \epsilon\, f'(x_0) + \frac{\epsilon^2}{2!} f''(x_0) + \frac{\epsilon^3}{3!} f'''(x_0) + \cdots \tag{6.3}$$

The special case of eqn (6.2) when $x_0 = 0$ is called a *Maclaurin series*. Apart from the origin, other common points of expansion for a Taylor series tend to be the locations of the maxima and minima of $f(x)$.

## 6.3 Some common examples

As a concrete example of the discussion above, let us work out the Maclaurin series for $\sin x$. Such problems are best done by setting up a two-column list with the function and its derivatives on the left, and their corresponding values at $x = x_0$ (in this case zero) on the right

$$
\begin{array}{llll}
f(x) = & \sin x & \therefore\ f(0) = & 0 \\
\therefore\ f'(x) = & \cos x & f'(0) = & 1 \\
f''(x) = & -\sin x & f''(0) = & 0 \\
f'''(x) = & -\cos x & f'''(0) = & -1
\end{array}
$$

Higher-order derivatives can be ascertained from this table by noting that the fourth one is $\sin x$, and then the cycle keeps repeating itself. Substituting for the various $f^n(0)$ in eqn (6.2), we obtain

$$\sin x = x - \frac{x^3}{3!} + \frac{x^5}{5!} - \frac{x^7}{7!} + \cdots \tag{6.4}$$

$\sin\left(\tfrac{1}{2}\right) = \tfrac{1}{2} - \tfrac{1}{48} + \tfrac{1}{3840} - \cdots$

This provides us with a means of calculating the sine of an angle without having to draw right-angled triangles; to improve the accuracy of the answer, we simply add more terms. The small angle approximation of eqn (3.7) follows immediately when $|x| \ll 1$, where $x$ is measured in radians, because the higher powers of $x$ rapidly become negligible.

A retracing of the steps of the analysis above for $f(x) = \cos x$ yields

$$\cos x = 1 - \frac{x^2}{2!} + \frac{x^4}{4!} - \frac{x^6}{6!} + \cdots \tag{6.5}$$

This expansion would also have resulted if we had differentiated, or integrated, eqn (6.4) term-by-term.

Finally, it is readily shown that the Taylor series of $f(x) = e^x$ is

$$e^x = 1 + x + \frac{x^2}{2!} + \frac{x^3}{3!} + \frac{x^4}{4!} + \frac{x^5}{5!} + \cdots \qquad (6.6)$$

Again, a term-by-term differentiation of eqn (6.6) indicates how the derivative of an exponential can be equal to the function itself.

$$e = 1 + 1 + \tfrac{1}{2} + \tfrac{1}{6} + \tfrac{1}{24} + \cdots$$

## 6.4 The radius of convergence

The power of a Taylor series lies in the fact that it enables us to approximate an arbitrary function by a simple low-order polynomial in the vicinity of a particular point; for greater accuracy, we just need to extend the expansion to a higher order. The last statement must be qualified in that its validity hinges on the property that the terms omitted from the summation become ever smaller and more negligible. This is only true as long as $|x - x_0| < R$, where the threshold $R$ is known as the *radius of convergence*; the Taylor series should not be used outside this domain. Formally, this convergence criterion requires that the ratio of neighbouring terms in the expansion, higher to lower, is less than unity at the infinite-order end of the series. While $R \to \infty$ for eqns (6.4)–(6.6), so that those three cases are valid for all values of $x$, the Taylor series for $f(x) = (1+x)^n$ has a very limited range of applicability

$$\lim_{n \to \infty} \left| \frac{a_{n+1} (x - x_0)^{n+1}}{a_n (x - x_0)^n} \right| < 1$$

$$(\text{for } a_n \neq 0 \text{ and } a_{n+1} \neq 0)$$

$$(1+x)^n = 1 + nx + \frac{n(n-1)x^2}{2!} + \frac{n(n-1)(n-2)x^3}{3!} + \cdots \qquad (6.7)$$

for $|x| < 1$. This is a generalized version of the binomial expansion met in section 1.5, where the power need not be a positive integer; if $n$ is a positive integer then the right-hand side of eqn (6.7) terminates (or becomes zero) after a finite number of terms, so that eqn (1.11) is recovered, and the expansion is valid for all values for $x$. Another Taylor series that is encountered frequently is

$$\sqrt{1+x} = 1 + \frac{x}{2} - \frac{x^2}{8} + \cdots$$

$$\ln(1+x) = x - \frac{x^2}{2} + \frac{x^3}{3} - \frac{x^4}{4} + \cdots \qquad (6.8)$$

which converges as long as $|x| < 1$.

## 6.5 l'Hôpital's rule

The Taylor series is a valuable tool for ascertaining limits in awkward situations. For example, the value of $\sin x / x$ is not obvious when $x = 0$; in essence, it is equivalent to trying to work out the product of nought and infinity! The desired ratio can be obtained by considering how the numerator and denominator behave as they gradually approach zero; in other words, by using a Taylor series expansion about the origin

$$\lim_{\delta x \to 0} \frac{\sin x}{x} = \lim_{\delta x \to 0} \frac{x - x^3/6 + \cdots}{x} = \lim_{\delta x \to 0} 1 - \frac{x^2}{6} + \cdots$$

the required result is seen to be unity.

Extending the arguments of the case above to the general ratio $g(x)/h(x)$, by expanding the numerator and the denominator in Taylor series about $x=a$, it is easily shown that

$$\lim_{x \to a} \frac{g(x)}{h(x)} = \lim_{x \to a} \frac{g'(x)}{h'(x)} \tag{6.9}$$

$$\lim_{x \to 0} \frac{\sin x}{x} = \lim_{x \to 0} \frac{\cos x}{1} = 1$$

if $g(a) = h(a) = 0$. If both the first derivatives are also zero at $x=a$, then the limit is given by the ratio of the second derivatives, $g''(a)/h''(a)$; and so on until a well-defined ratio is obtained. This is known as *l'Hôpital's rule*.

## 6.6 Newton-Raphson algorithm

The Taylor series also provides a powerful means of finding the roots of an equation numerically; that is to say, ascertaining the values of $x$ which satisfy the relationship $f(x) = 0$. For an arbitrary function, $f(x)$, the solutions may be difficult to obtain algebraically. Nevertheless we may know that $x = x_0$ is a good guess, so that $f(x_0) \approx 0$; how can we use this to derive a better estimate? Well, expanding $f(x)$ in a Taylor series about $x = x_0$ we have

$$f(x) = f(x_0) + (x - x_0) f'(x_0) + \cdots = 0$$

Assuming that our initial guess is good enough that the quadratic and higher-order terms are negligible by comparison with the zeroth and first-order ones, the resulting linear equations can readily be solved to yield

$$x \approx x_0 - \frac{f(x_0)}{f'(x_0)} \tag{6.10}$$

In other words, the values of the function and its first derivative at $x_0$ enable us to obtain a better estimate of the root according to eqn (6.10). This can, of course, be used as our new initial guess, and the procedure repeated until there is no significant change between $x$ and $x_0$; this *iterative* numerical method of solving $f(x) = 0$ is known as the *Newton-Raphson* algorithm.

## 6.7 Worked examples

(1) Derive the Taylor series for $\ln(1+x)$ and $\ln(1-x)$. Hence, write down the power series for $\ln\left[(1+x)/(1-x)\right]$.

$$\text{Let} \quad f(x) = \ln x \qquad\qquad \therefore \ f(1) = 0$$
$$\therefore \ f'(x) = x^{-1} \qquad\qquad f'(1) = 1$$
$$f''(x) = -x^{-2} \qquad\qquad f''(1) = -1$$
$$f'''(x) = 2x^{-3} \qquad\qquad f'''(1) = 2$$

$$f''''(x) = -6x^{-4} \qquad\qquad f''''(1) = -6$$
$$f'''''(x) = 24x^{-5} \qquad\qquad f'''''(1) = 24$$

But $\quad f(1+x) = f(1) + x\,f'(1) + \dfrac{x^2}{2!}f''(1) + \dfrac{x^3}{3!}f'''(1) + \cdots$

$\therefore \quad \ln(1+x) = x - \dfrac{x^2}{2} + \dfrac{x^3}{3} - \dfrac{x^4}{4} + \dfrac{x^5}{5} - \cdots \quad$ for $|x|<1$

$\displaystyle\lim_{n\to\infty}\left|\dfrac{nx}{n+1}\right| < 1$

for convergence

If we substitute $x = -u$ (say) in the above, then it follows that

$$\ln(1-x) = -x - \dfrac{x^2}{2} - \dfrac{x^3}{3} - \dfrac{x^4}{4} - \dfrac{x^5}{5} - \cdots \quad \text{for } |x|<1$$

$$\ln\left(\dfrac{1+x}{1-x}\right) = \ln(1+x) - \ln(1-x)$$

$$= 2\left(x + \dfrac{x^3}{3} + \dfrac{x^5}{5} + \cdots\right) \quad \text{for } |x|<1$$

---

**(2)** Derive the general binomial expansion for $(1+x)^n$. By expressing $\sqrt{A}$ as $\alpha(1+x)^{1/2}$, where $\alpha$ and $x$ are suitable constants, find $\sqrt{8}$ and $\sqrt{17}$ to four decimal places.

---

Let $\quad f(x) = x^n \qquad\qquad \therefore\ f(1) = 1$

$\therefore\ f'(x) = n\,x^{n-1} \qquad\qquad f'(1) = n$

$\quad f''(x) = n(n-1)x^{n-2} \qquad f''(1) = n(n-1)$

$\quad f'''(x) = n(n-1)(n-2)x^{n-3} \qquad f'''(1) = n(n-1)(n-2)$

But $\quad f(1+x) = f(1) + x\,f'(1) + \dfrac{x^2}{2!}f''(1) + \dfrac{x^3}{3!}f'''(1) + \cdots$

$\therefore\ (1+x)^n = 1 + nx + \dfrac{n(n-1)}{2}x^2 + \dfrac{n(n-1)(n-2)}{6}x^3 + \cdots \quad$ for $|x|<1$

$\displaystyle\lim_{k\to\infty}\left|\left(\dfrac{n-k}{k+1}\right)x\right| < 1$

(if $n$ not a positive integer)

$$\sqrt{8} = \sqrt{9-1} = 3(1+x)^{1/2} \quad \text{where } x = -1/9$$

But $\quad n = \tfrac{1}{2} \Rightarrow \sqrt{1+x} = 1 + \tfrac{1}{2}x - \tfrac{1}{8}x^2 + \tfrac{1}{16}x^3 - \tfrac{5}{128}x^4 + \cdots$

$\therefore\ \sqrt{8} = 3\left[1 - \tfrac{1}{18} - \tfrac{1}{648} - \tfrac{1}{11664} - \tfrac{5}{839808} - \cdots\right]$

$\qquad = \underline{2.8284} \quad$ to 4 decimal places

$$\sqrt{17} = \sqrt{16+1} = 4\,(1+x)^{1/2} \quad \text{where} \quad x = 1/16$$

$$\therefore \ \sqrt{17} = 4\left[1 + \tfrac{1}{32} - \tfrac{1}{2048} + \tfrac{1}{65536} - \cdots\right]$$

$$= \underline{4.1231} \quad \text{to 4 decimal places}$$

The binomial expansion converges increasingly quickly as $|x| \to 0$. Thus, for example, in the calculation of $\sqrt{1.000001} = (1+10^{-6})^{1/2}$, the inclusion of the $x^2$ term (which is $1.25 \times 10^{-13}$) yields greater accuracy than that provided by most electronic calculators! More seriously, though, a sum like $\sqrt{a^2+b^2}$ where $a \gg b$ is best done by using the binomial expansion

$$\sqrt{a^2+b^2} = a\left[1 + \left(\tfrac{b}{a}\right)^2\right]^{1/2} = a\left[1 + \tfrac{1}{2}\left(\tfrac{b}{a}\right)^2 - \tfrac{1}{8}\left(\tfrac{b}{a}\right)^4 + \tfrac{1}{16}\left(\tfrac{b}{a}\right)^6 - \cdots\right]$$

to avoid 'rounding errors', even in high precision computers, when $|b/a| \ll 1$.

---

**(3)** Derive the Maclaurin series for $\cos^{-1} x$.

---

The Maclaurin series is simply a special case of a Taylor series where the expansion is about the origin. We could ascertain it in the usual systematic fashion by setting up a two-column list with $f(x) = \cos^{-1} x$ and its derivatives on the left, and their values at $x = 0$ on the right. An algebraically much easier way, however, is to differentiate $\cos^{-1} x$ just once, use a binomial expansion and integrate term-by-term.

$$\text{Let} \quad f(x) = \cos^{-1} x \qquad \therefore \ f'(x) = \frac{-1}{\sqrt{1-x^2}} = -\left(1-x^2\right)^{-1/2}$$

$$\text{But} \quad \left(1-x^2\right)^{-1/2} = 1 + \tfrac{1}{2}x^2 + \tfrac{3}{8}x^4 + \tfrac{5}{16}x^6 + \tfrac{35}{128}x^8 + \cdots$$

$$\therefore \ f(x) = \int f'(x)\,\mathrm{d}x$$

$$= K - x - \tfrac{1}{6}x^3 - \tfrac{3}{40}x^5 - \tfrac{5}{112}x^7 - \tfrac{35}{1152}x^9 - \cdots$$

$$\text{But} \quad \cos^{-1}(0) = \tfrac{\pi}{2} \ \Rightarrow \ K = \tfrac{\pi}{2}$$

$$\therefore \ \underline{\cos^{-1} x = \tfrac{\pi}{2} - x - \tfrac{1}{6}x^3 - \tfrac{3}{40}x^5 - \tfrac{5}{112}x^7 - \cdots}$$

We should remember that, like $\mathrm{d}/\mathrm{d}x\,(\cos^{-1} x) = -1/\sqrt{1-x^2}$, the Maclaurin series is only valid for $|x| < 1$ and $0 \leqslant \cos^{-1} x \leqslant \pi$.

**(4)** Determine the following limits:

(a) $\displaystyle\lim_{x\to 0}\frac{\sin(ax)}{x}$    (b) $\displaystyle\lim_{x\to 0}\frac{\cos x - 1}{x}$    (c) $\displaystyle\lim_{x\to 0}\frac{2\cos x + x\sin x - 2}{x^4}$

(a)  $\displaystyle\lim_{x\to 0}\frac{\sin(ax)}{x} = \lim_{x\to 0}\frac{ax - \frac{1}{6}a^3 x^3 + \cdots}{x}$

$\displaystyle\qquad\qquad = \lim_{x\to 0}\ a - \frac{1}{6}a^3 x^2 + \cdots = \underline{a}$

Alternatively, we could use l'Hôpital's rule:

$\displaystyle\lim_{x\to 0}\frac{\sin(ax)}{x} = \lim_{x\to 0}\frac{a\cos(ax)}{1} = \underline{a}$

(b)  $\displaystyle\lim_{x\to 0}\frac{\cos x - 1}{x} = \lim_{x\to 0}\frac{\left(1 - \frac{1}{2}x^2 + \frac{1}{24}x^4 - \cdots\right) - 1}{x}$

$\displaystyle\qquad\qquad = \lim_{x\to 0}\ -\frac{1}{2}x + \frac{1}{24}x^3 - \cdots = \underline{0}$

or  $\displaystyle\lim_{x\to 0}\frac{\cos x - 1}{x} = \lim_{x\to 0}\frac{-\sin x}{1} = \underline{0}$    (l'Hôpital)

(c)  $\displaystyle\lim_{x\to 0}\frac{2\cos x + x\sin x - 2}{x^4}$

$\displaystyle\qquad = \lim_{x\to 0}\frac{2\left(1 - \frac{1}{2}x^2 + \frac{1}{24}x^4 - \cdots\right) + x\left(x - \frac{1}{6}x^3 + \cdots\right) - 2}{x^4}$

$\displaystyle\qquad = \lim_{x\to 0}\ -\frac{1}{12} + \mathcal{O}(x^2) = \underline{-\frac{1}{12}}$

The notation $\mathcal{O}(x^2)$ is read as 'terms of order $x^2$'; that is to say, the omitted terms contain a factor of $x^2$ and other contributions with powers of $x$ greater than two.

or  $\displaystyle\lim_{x\to 0}\frac{2\cos x + x\sin x - 2}{x^4} = \lim_{x\to 0}\frac{-\sin x + x\cos x}{4x^3}$    (l'Hôpital)

$\displaystyle\qquad\qquad = \lim_{x\to 0}\frac{-x\sin x}{12x^2}$    (l'Hôpital)

$\displaystyle\qquad\qquad = \lim_{x\to 0}\frac{-\sin x}{12x}$

$\displaystyle\qquad\qquad = \lim_{x\to 0}\frac{-\cos x}{12}$    (l'Hôpital)

$\displaystyle\qquad\qquad = \underline{-\frac{1}{12}}$

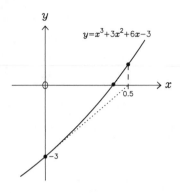

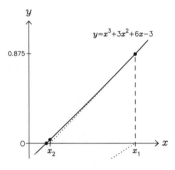

**(5)** Given that $x^3 + 3x^2 + 6x - 3 = 0$ has only one (real) root, find its value to five significant figures by using the Newton-Raphson method.

Let   $f(x) = x^3 + 3x^2 + 6x - 3$    $\therefore$  $f'(x) = 3x^2 + 6x + 6$

If $f(x_n) \approx 0$, then a better estimate of $f(x) = 0$, $x_{n+1}$, is given by

$$x_{n+1} = x_n - \frac{f(x_n)}{f'(x_n)} = x_n - \frac{x_n^3 + 3x_n^2 + 6x_n - 3}{3x_n^2 + 6x_n + 6}$$

Let   $x_0 = 0$                    $\therefore$  $f(x_0) = -3$

$\therefore$  $x_1 = 0.5$                    $f(x_1) = 0.875$

$x_2 = 0.4102564$            $f(x_2) = 0.03552$

$x_3 = 0.4062950$            $f(x_3) = 0.00006633$

$x_4 = 0.4062876$            $f(x_4) = 0$

i.e.   $x^3 + 3x^2 + 6x - 3 = 0$   when   $\underline{x = 0.40629}$   (5 sig. fig.)

---

**(6)** Einstein's famous formula, $E = mc^2$, which relates a mass $m$ to energy $E$ through the speed of light $c$, generalizes to $E = mc^2/\sqrt{1 - (v/c)^2}$ when the object is travelling at speed $v$. Show that at non-relativistic speeds, $v \ll c$, this amounts to the addition of kinetic energy $\frac{1}{2}mv^2$.

$$E = \frac{mc^2}{\sqrt{1 - \left(\frac{v}{c}\right)^2}} = mc^2\left[1 - \left(\frac{v}{c}\right)^2\right]^{-1/2}$$

$$= mc^2\left[1 + \frac{1}{2}\left(\frac{v}{c}\right)^2 + \frac{3}{8}\left(\frac{v}{c}\right)^4 + \cdots\right]$$

$$= mc^2 + \frac{1}{2}mv^2\left[1 + \mathcal{O}\left(\left(\frac{v}{c}\right)^2\right)\right]$$

# 7 Complex numbers

## 7.1 Definition

If any number, integer or fraction, positive or negative, is multiplied by itself, then the result is always greater than, or equal to, zero. What, then, is the square root of $-9$? To address this question we need to invent an *imaginary* number, usually denoted by 'i', whose square is defined to be negative

$$i^2 = -1 \qquad (7.1)$$

A *real* number, say $b$ (where $b^2 \geqslant 0$), times i is also imaginary; it's just $b$ times bigger than i. If $a$ is also an ordinary number, then the sum $z$ of $a$ and $ib$

$$z = a + ib \qquad (7.2)$$

is known as a *complex* number; this does not indicate an intrinsic difficulty with the concept, but highlights the hybrid nature of the entity. It consists of both a real part and an imaginary one

$$\mathcal{Re}\{z\} = a \quad \text{and} \quad \mathcal{Im}\{z\} = b \qquad (7.3)$$

It may seem odd that $\mathcal{Im}\{z\}$ is $b$ rather than $ib$, but this is so because it represents the size of the imaginary component.

Although the construct of imaginary numbers appears arbitrary, complex numbers turn out to be very useful for tackling many real-life problems. Our aims here, however, are to acquire a knowledge of their basic properties.

## 7.2 Basic algebra

Let us start with the most elementary operations, namely addition and subtraction. To add, or subtract, complex numbers, we simply combine the real and imaginary parts separately. For example

$$a + ib \pm (c + id) = a \pm c + i(b \pm d) \qquad (7.4)$$

where $a$, $b$, $c$, and $d$ are real. In fact we have made an implicit assumption in writing the term on the far right, in that complex numbers obey the same rules

$$1 + 2i - (5 - i) = -4 + 3i$$

as ordinary ones; but with every occurrence of $i^2$ replaced by $-1$. Thus, it is easy to show that the product of $a+ib$ and $c+id$ is given by

$$(a+ib)(c+id) = ac - bd + i(ad + bc) \qquad (7.5)$$

$$(1+2i)(3-i) = 5+5i$$

because $i^2bd = -bd$. Division is less straightforward in that it entails the use of the *complex conjugate* of the denominator; let us consider this first.

The conjugate of a complex number $z$, denoted by $z^*$, is defined to have the same real part but the opposite imaginary component: $\mathcal{Re}\{z^*\} = \mathcal{Re}\{z\}$ and $\mathcal{Im}\{z^*\} = -\mathcal{Im}\{z\}$. In terms of eqn (7.2), therefore

$$z^* = (a+ib)^* = a - ib \qquad (7.6)$$

Thus, complex numbers and their conjugates satisfy the following relationships

$$
\begin{aligned}
z + z^* &= & 2a & = 2\mathcal{Re}\{z\} \\
z - z^* &= & 2bi & = 2\mathcal{Im}\{z\}i \\
z\,z^* &= a^2 + b^2 & = |z|^2
\end{aligned}
\qquad (7.7)
$$

We will come to the meaning of $|z|$ shortly, but the important point about eqn (7.7) is that the product $z\,z^* = a^2 + b^2$ is a real number because it does not involve any $i$'s. This feature enables us to calculate the real and imaginary part of the ratio of two complex numbers by multiplying both the top and bottom of the quotient by the conjugate of the denominator

$$\frac{a+ib}{c+id} = \frac{a+ib}{c+id} \times \frac{c-id}{c-id} = \frac{ac + bd + i(bc - ad)}{c^2 + d^2} \qquad (7.8)$$

To evaluate the ratio $(1+2i)/(3-i)$, for example, we multiply it by unity in the form $(3+i)/(3+i)$; this gives a real denominator of $3^2 + 1^2 = 10$ and a complex numerator of $1 + 7i$. Hence the result is $1/10 + (7/10)i$.

## 7.3 The Argand diagram

So far we have considered complex numbers from an algebraic point-of-view; it is often helpful to think of them in geometrical terms. This is easily done with the aid of an *Argand diagram* where the horizontal, or $x$, axis of a graph is seen as representing the real part of a complex number, and the vertical, or $y$, axis gives the imaginary component. Thus the point with $(x,y)$ coordinates $(a,b)$ corresponds to the complex number $z = a+ib$. Its conjugate $z^*$ is a reflection in the real axis. An alternative way of specifying the location of a point on a graph is through its distance $r$ from the origin, and the anticlockwise angle $\theta$ that this 'radius' makes with the (positive) real axis. In this system $r$ is known as the *modulus, magnitude,* or *amplitude* of $z$; $\theta$ is called its *argument* or *phase*.

By using the elementary trigonometry met in section 3.2, we can relate the quantities $r$, $\theta$, $a$, and $b$ in the Argand diagram by

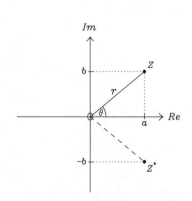

$$a = r\cos\theta \quad \text{and} \quad b = r\sin\theta \tag{7.9}$$

or, in the reverse sense, through

$$r^2 = a^2 + b^2 \quad \text{and} \quad \theta = \tan^{-1}\left(\tfrac{b}{a}\right) \tag{7.10}$$

A comparison of eqns (7.7) and (7.10) shows that $z\,z^* = r^2$, where $r = |z|$ is the modulus of the complex number. We need to qualify the second part of eqn (7.10) in that there is ambiguity in $\tan^{-1}(b/a)$. If $z = -1-i$, for example, then a plot of this point in the Argand diagram shows $\theta$ to be $5\pi/4$ radians or $225°$; although this is a value of $\tan^{-1}(1)$, so too is $45°$ which does not give the correct argument. It is also worth remembering that $\theta$ is only defined to within a factor of $2\pi$, because we could add (or subtract) any integer number of $360°$ to it and obtain the same point in the Argand diagram.

## 7.4 The imaginary exponential

Perhaps the single most important result in complex analysis concerns the exponential of an imaginary number

$$e^{i\theta} = \cos\theta + i\sin\theta \tag{7.11}$$

While this equation is not immediately obvious, it is easily verified by substituting $x = i\theta$ in the Taylor series expansion of eqn (6.6) and collecting together the odd and even powers of $\theta$ separately; a comparison with eqns (6.4) and (6.5), while remembering that $i^2 = -1$, then yields eqn (7.11). The product of $r$ and $e^{i\theta}$ allows a complex number to be expressed in a very compact form in terms of its modulus and argument:

$$
\begin{aligned}
e^{i\pi/4} &= \tfrac{1}{\sqrt{2}}(1+i)\\
e^{i\pi/2} &= i\\
e^{i3\pi/4} &= \tfrac{1}{\sqrt{2}}(-1+i)\\
e^{i\pi} &= -1
\end{aligned}
$$

$$a + ib = r(\cos\theta + i\sin\theta) = r\,e^{i\theta} \tag{7.12}$$

where $a$, $b$, $r$, and $\theta$ are related through eqns (7.9) and (7.10).

As we shall see shortly, the exponential form of a complex number is very useful when dealing with roots and logarithms; it also provides a valuable insight into products and quotients. Consider two complex numbers $z_1$ and $z_2$ having amplitudes $r_1$ and $r_2$, and phases $\theta_1$ and $\theta_2$

$$z_1 = r_1\,e^{i\theta_1} \quad \text{and} \quad z_2 = r_2\,e^{i\theta_2}$$

Then, using the rules for combining indices in section 1.2, we obtain

$$z_1 z_2 = r_1 r_2\,e^{i(\theta_1+\theta_2)} \tag{7.13}$$

In other words, by comparing the right-hand side with the standard form $r\,e^{i\theta}$, we see that the modulus of a product is equal to the product of the moduli; and that the argument of a product is the sum of the arguments. Similarly, the modulus of a quotient is equal to the ratio of the moduli, and its argument is the difference of the arguments

$$\frac{z_1}{z_2} = \frac{r_1}{r_2}\, e^{i(\theta_1 - \theta_2)} \qquad (7.14)$$

The product of a complex number with its conjugate is a special case of eqn (7.13) with $z_2 = z_1^*$, so that $r_2 = r_1$ and $\theta_2 = -\theta_1$, and immediately returns the now familiar result that $z\,z^* = |z|^2$.

## 7.5  Roots and logarithms

At the beginning of this chapter we posed the question of $\sqrt{-9}$. While the answer to this simple problem may now be intuitively obvious, as $\pm 3i$, how can we work it out in a systematic fashion? Well, technically, we wish to find the value of the complex number $z$ which satisfies the equation $z^2 = -9$. This task becomes easier if we write the right-hand side in the form $r\,e^{i\theta}$. For $-9$, the amplitude is 9 and the phase is $180°$, give or take any integer multiple of $360°$; thus we have

$$z^2 = -9 = 9\, e^{i(\pi \pm 2\pi n)}$$

where $n$ is any integer (and $\theta$ is in radians). Raising both sides to the power of a half, according to the rules in section 1.2, gives the square root as

$$z = 9^{1/2}\, e^{i(\pi \pm 2\pi n)/2} = 3\, e^{i(\pi/2 \pm n\pi)}$$

Hence, an even value of $n$ (such as $n = 0$) yields the solution $z = 3i$, whereas an odd $n$ (like $n = 1$) gives the alternative result of $z = -3i$.

As a second example of this type of calculation, let's consider the cube roots of $i$; that is, the solution of the equation $z^3 = i$. Again, the best procedure is to write the right-hand side in the form $r\,e^{i\theta}$

$$z^3 = i = e^{i(\pi/2 \pm 2\pi n)}$$

where $n$ is any integer. Raising both sides to the power of a third, we obtain

$$z = e^{i(\pi/2 \pm 2\pi n)/3} = e^{i(\pi/6 \pm n2\pi/3)}$$

Three distinct solutions emerge for $n = 0$, $n = 1$, and $n = 2$; namely $\frac{1}{2}(\sqrt{3} + i)$, $\frac{1}{2}(-\sqrt{3} + i)$, and $-i$ respectively. All other values of $n$ keep generating the same results (e.g. $n = 3$ is identical to $n = 0$).

If the cube roots of $i$ are plotted on an Argand diagram, they are seen to lie at the vertices of an equilateral triangle. Such a uniform distribution of solutions around the circumference of a circle is a general feature of this type of problem, with the $m$-roots of a complex number leading to an $m$-sided regular polygon.

In addition to its use for roots and powers, the $r\,e^{i\theta}$ form of a complex number allows us to extend the domain of logarithms. When restricted to real numbers,

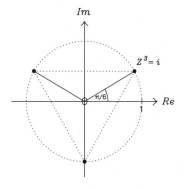

a logarithm could only be defined for positive quantities; now, using the rules of section 1.2, we can take the logarithm of anything

$$\ln(r e^{i\theta}) = \ln(r) + i\theta \tag{7.15}$$

The amplitude $r$ is positive by definition, and the phase $\theta$ is usually taken as being zero for real numbers. Since we know that $\theta$ entails an ambiguity with respect to the addition of integer numbers of $2\pi$ radians, the imaginary part of a logarithm can take an infinite, but discrete, number of values; by convention, the *principal* one is chosen to lie between $-\pi$ and $+\pi$ radians.

$$\ln(2i) = \ln\left[2 e^{i(\pi/2 \pm 2\pi n)}\right]$$
$$= \ln 2 + i\left(\tfrac{\pi}{2} \pm 2\pi n\right)$$

## 7.6 De Moivre's theorem and trigonometry

A useful trigonometric result also follows from the imaginary exponential of section 7.4 when it is combined with eqn (1.5)

$$(e^{i\theta})^m = e^{im\theta}$$

Using the definition of eqn (7.11) on both sides separately, we obtain *de Moivre's theorem*

$$(\cos\theta + i\sin\theta)^m = \cos(m\theta) + i\sin(m\theta) \tag{7.16}$$

This yields, for example, the formulae for the sines and cosines of multiple angles. As a simple illustration, consider the case of $m=2$

$$\cos^2\theta + 2i\sin\theta\cos\theta - \sin^2\theta = \cos(2\theta) + i\sin(2\theta)$$

which, on equating the real and imaginary parts, returns the double-angle formulae of eqns (3.14) and (3.11) respectively.

Further connections between the imaginary exponential and trigonometry emerge if we combine eqn (7.11) with its complex conjugate

$$e^{-i\theta} = \cos\theta - i\sin\theta$$

Their sum gives a formula for $\cos\theta$, while the difference yields $\sin\theta$

$$\sin\theta = \frac{e^{i\theta} - e^{-i\theta}}{2i} \quad \text{and} \quad \cos\theta = \frac{e^{i\theta} + e^{-i\theta}}{2} \tag{7.17}$$

The tangent is obviously given by the ratio of these expressions, as defined in eqn (3.3). Amongst other things, these relationships are useful for rewriting powers of sines and cosines in terms of their mutiple-angle equivalents. To take a very simple example

$$\sin^2\theta = \left(\frac{e^{i\theta} - e^{-i\theta}}{2i}\right)^2 = \frac{e^{i2\theta} - 2 + e^{-i2\theta}}{-4} = \frac{1}{2}\left[1 - \left(\frac{e^{i2\theta} + e^{-i2\theta}}{2}\right)\right]$$

which can be recognized as being $(1 - \cos 2\theta)/2$, as in eqn (3.15).

## 7.7 Hyperbolic functions

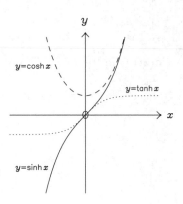

All the trigonometric functions we met in chapter 3 have a set of counterparts known as *hyperbolic* (trigonometric) functions; they are designated by having an 'h' appended to the abbreviations of the former. For example, sinh, cosh, and tanh are the hyperbolic sin, cos, and tan respectively. We will see the relevance to complex numbers shortly, but begin with the definition of $\sinh x$ and $\cosh x$

$$\sinh x = \frac{e^x - e^{-x}}{2} \quad \text{and} \quad \cosh x = \frac{e^x + e^{-x}}{2} \tag{7.18}$$

By analogy with eqn (3.3), the hyperbolic tangent is the ratio of sinh and cosh

$$\tanh x = \frac{\sinh x}{\cosh x} = \frac{e^x - e^{-x}}{e^x + e^{-x}} = \frac{e^{2x} - 1}{e^{2x} + 1} \tag{7.19}$$

The shapes of the curves $y = \sinh x$, $\cosh x$, and $\tanh x$ can be ascertained by considering the characteristics of the exponential function met in section 2.4; in particular, by looking at the behaviour of $y = e^x$ and $y = e^{-x}$ in the limits of $x \to \pm\infty$ and at $x = 0$.

The relationship between the hyperbolic and ordinary (circular) trigonometric functions becomes obvious if we substitute $\theta = ix$ into eqn (7.17) and compare the result with eqn (7.18); it yields

$$\sin(ix) = i\sinh(x) \quad \text{and} \quad \cos(ix) = \cosh(x) \tag{7.20}$$

In other words, the hyperbolic functions are related to the sines, cosines, and so on, of imaginary angles. By using eqn (7.20), we can derive identities for hyperbolic functions which are equivalent to those discussed in section 3.3. For example, since $\sin^2(ix) + \cos^2(ix) = 1$, according to eqn (3.8), we have

$$\cosh^2 x - \sinh^2 x = 1 \tag{7.21}$$

$$1 - \tanh^2 x = \text{sech}^2 x$$

$$1 - \coth^2 x = -\text{cosech}^2 x$$

Indeed, this sort of manipulation leads to an observation, known as *Osborne's rule*, that the Pythagorean identities for hyperbolic functions are the same as those for trigonometric ones except that the sign of every $\sinh^2 x$, or implied $\sinh^2 x$ (as in $\tanh^2 x = \sinh^2 x / \cosh^2 x$) is reversed.

The derivatives, and integrals, of hyperbolic functions are also readily ascertained by exploiting the convenient differential properties of the exponential function. Thus, using the definition of eqn (7.18), and the material in chapter 4, it is easily shown that

$$\frac{d}{dx}(\sinh x) = \cosh x \quad \text{and} \quad \frac{d}{dx}(\cosh x) = \sinh x \tag{7.22}$$

$$\frac{d}{dx}(\tanh x) = \text{sech}^2 x$$

and so on. The derivatives of the inverse hyperbolic functions can be obtained in an analogous manner to those of the inverse trigonometric functions. That is to say, by differentiating $x = \sinh y$ with respect to $x$, and then using eqn (7.21), to yield $dy/dx$ where $y = \sinh^{-1} x$

$$\frac{d}{dx}\left(\sinh^{-1}x\right) = \frac{1}{\sqrt{1+x^2}} \tag{7.23}$$

Incidentally, explicit formulae for the inverse hyperbolic functions, such as

$$\frac{d}{dx}\left(\cosh^{-1}x\right) = \frac{1}{\sqrt{x^2-1}}$$

$$\tanh^{-1}x = \frac{1}{2}\ln\left(\frac{1+x}{1-x}\right) \tag{7.24}$$

where $|x| \leqslant 1$, can be derived from the exponential definitions of eqns (7.18) and (7.19). The derivative of $\tanh^{-1}x$ can then be confirmed by differentiating eqn (7.24) with respect to $x$.

$$\frac{d}{dx}\left(\tanh^{-1}x\right) = \frac{1}{1-x^2}$$

## 7.8 Some useful properties

We mentioned earlier that complex numbers provide a powerful analytical tool for tackling many advanced theoretical problems. While our examples here are necessarily of an elementary nature, we should specifically state the properties of complex numbers that often make them so useful. Namely, the exponential decomposition of eqn (7.11) and the linearity with respect to addition and subtraction. That is to say 'the sum of the real parts is equal to the real part of the sum', or put mathematically

$$\sum_{k=1}^{N} \mathcal{R}e\{z_k\} = \mathcal{R}e\left\{\sum_{k=1}^{N} z_k\right\} \tag{7.25}$$

where $z_k$ is the $k^{th}$ complex number, and $k = 1, 2, 3, \ldots, N$. The same is true of the imaginary components, of course, and differences. If $z$ is a complex number which is a function of $t$, say, such as $t^2 + i\cos(t)$, then the linearity of eqn (7.4) leads to the properties

$$\sum_{k=0}^{\infty} \frac{\sin(k\theta)}{k!} = \sum_{k=0}^{\infty} \mathcal{I}m\left\{\frac{e^{ik\theta}}{k!}\right\}$$

$$= \mathcal{I}m\left\{\sum_{k=0}^{\infty} \frac{(e^{i\theta})^k}{k!}\right\}$$

$$\frac{d}{dt}\left(\mathcal{R}e\{z\}\right) = \mathcal{R}e\left\{\frac{dz}{dt}\right\} \quad\text{and}\quad \int \mathcal{R}e\{z\}\,dt = \mathcal{R}e\left\{\int z\,dt\right\} \tag{7.26}$$

Again, the same goes for the imaginary parts.

Equations (7.25) and (7.26) are useful because their right-hand sides are frequently easier to evaluate than the corresponding expressions on the left. Let us illustrate this with an example involving integration

$$I = \int e^{at}\cos(bt)\,dt \tag{7.27}$$

where $a$ and $b$ are real. This problem can be solved by using integration by parts twice, as in section 5.6, or by using complex numbers. The latter requires us to spot that the integrand can be written in the form

$$e^{at}\cos(bt) = e^{at}\mathcal{R}e\{e^{ibt}\} = \mathcal{R}e\{e^{(a+ib)t}\} \tag{7.28}$$

Hence, on using the property of eqn (7.26), we are left with an easy integral

$$I = \mathcal{R}e\left\{\int e^{(a+ib)t}\,dt\right\} = \mathcal{R}e\left\{\frac{e^{(a+ib)t}}{a+ib}\right\} + K \qquad (7.29)$$

where $K$, the constant of integration, is real. While a little effort is required to ascertain the real part of the quotient in eqn (7.29), starting with multiplication of the top and bottom by the complex conjugate of the denominator, $(a-ib)$, it is not very difficult to show that

$$\int e^{at}\cos(bt)\,dt = \frac{e^{at}}{a^2+b^2}\Big[a\cos(bt) + b\sin(bt)\Big] + K$$

or to notice that the imaginary part yields the integral of $e^{at}\sin(bt)$.

## 7.9 Worked examples

---

**(1)** If $u = 2+3i$ and $v = 1-i$, find the real and imaginary parts of
(a) $u+v$, (b) $u-v$, (c) $uv$, (d) $u/v$ and (e) $v/u$.

---

(a) $u+v = 2+1 + i(3-1) = \underline{3+2i}$

(b) $u-v = 2-1 + i(3+1) = \underline{1+4i}$

(c) $uv = (2+3i)(1-i) = 2-2i+3i-3i^2 = \underline{5+i}$

(d) $\dfrac{u}{v} = \dfrac{2+3i}{1-i} = \left(\dfrac{2+3i}{1-i}\right)\times\left(\dfrac{1+i}{1+i}\right) = \dfrac{2+2i+3i+3i^2}{1+i-i-i^2} = \dfrac{-1+5i}{\underline{2}}$

(e) $\dfrac{v}{u} = \left(\dfrac{1-i}{2+3i}\right)\times\left(\dfrac{2-3i}{2-3i}\right) = \dfrac{2-3i-2i+3i^2}{4+9} = \dfrac{-1-5i}{\underline{13}}$

or $= \dfrac{1}{u/v} = \left(\dfrac{2}{-1+5i}\right)\times\left(\dfrac{-1-5i}{-1-5i}\right) = \dfrac{-2-10i}{1+25} = \dfrac{-1-5i}{\underline{13}}$

---

**(2)** For the previous example, evaluate:
(a) $|u|$, (b) $|v|$, (c) $|uv|$, (d) $|u/v|$ and (e) $|v/u|$.

---

(a) $|u|^2 = uu^* = (2+3i)(2-3i) = 4+9$     $\therefore\ |u| = \sqrt{13}$

(b) $|v|^2 = vv^* = (1-i)(1+i) = 1+1$     $\therefore\ |v| = \sqrt{2}$

(c) $|uv|^2 = (uv)(uv)^* = (5+i)(5-i) = 25+1$     $\therefore\ |uv| = \sqrt{26}$

This confirms the general result that the 'modulus of a product is equal to the product of the moduli'.

(d) $\left|\dfrac{u}{v}\right|^2 = \left[\dfrac{u}{v}\right]\left[\dfrac{u}{v}\right]^* = \left(\dfrac{-1+5i}{2}\right)\left(\dfrac{-1-5i}{2}\right) = \dfrac{1+25}{4}$    $\therefore$  $\left|\dfrac{u}{v}\right| = \sqrt{\dfrac{13}{2}}$

Again this verifies the general result that the 'modulus of a quotient is equal to the ratio of the moduli'.

(e) $\left|\dfrac{v}{u}\right|^2 = \left[\dfrac{v}{u}\right]\left[\dfrac{v}{u}\right]^* = \left(\dfrac{-1-5i}{13}\right)\left(\dfrac{-1+5i}{13}\right) = \dfrac{1+25}{169}$    $\therefore$  $\left|\dfrac{v}{u}\right| = \sqrt{\dfrac{2}{13}}$

This confirms the property that $|1/z| = 1/|z|$, for any complex number $z$; or, in this specific case, that $|v/u| = 1/|u/v|$.

---

**(3)** If $z = 1 + i\sqrt{3}$, sketch the following in an Argand diagram:
  $z$, $z^*$, $z^2$, $z^3$, $iz$, and $1/z$.

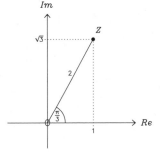

$$z = 1 + i\sqrt{3} = 2\,e^{i\pi/3}$$

That is to say, $z$ has a modulus, or amplitude, of $2$ and an argument, or phase, of $60°$.

$\therefore$  $z^* = 2\,e^{-i\pi/3}$,    $z^2 = 4\,e^{i2\pi/3}$   and   $z^3 = 8\,e^{i\pi} = -8$

$iz = e^{i\pi/2}\,2\,e^{i\pi/3} = 2\,e^{i(\pi/3 + \pi/2)} = 2\,e^{i5\pi/6}$   and   $\dfrac{1}{z} = \dfrac{1}{2}\,e^{-i\pi/3}$

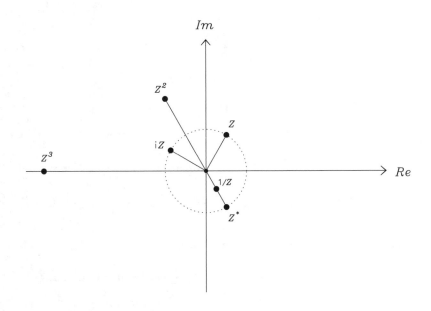

---

**(4)** Solve the quadratic equation $z^2 - z + 1 = 0$.

---

If $az^2 + bz + c = 0$, then $z = \dfrac{-b \pm \sqrt{b^2 - 4ac}}{2a}$

$z^2 - z + 1 = 0 \Rightarrow a = 1, \ b = -1 \ \text{and} \ c = 1$

$$\therefore \quad z = \tfrac{1 \pm \sqrt{1-4}}{2} = \tfrac{1}{2}\left(1 \pm i\sqrt{3}\right)$$

---

**(5)** Solve the following equations:
(a) $z^5 = 1$, (b) $z^5 = 1 + i$, (c) $(z+1)^5 = 1$ and (d) $(z+1)^5 = z^5$.
Sketch the solutions for (a) on an Argand diagram.

---

(a) $\quad z^5 = 1 = e^{i2\pi n}$ where $n = 0, \pm 1, \pm 2, \pm 3, \ldots$

$$\therefore \quad \underline{z = e^{in2\pi/5}} \qquad \text{for} \quad n = 0, 1, 2, 3, 4$$

Although the result for $z$ holds for any integer $n$, only 5 distinct solutions emerge; an alternative choice for $n$ which would specify them is $n = 0, \pm 1, \pm 2$. The appearance of 5 solutions was to be expected, as 'an $n^{\text{th}}$-order polynomial has exactly $n$ roots'; they are either real, or come as complex conjugate pairs (as in exercise 4 above).

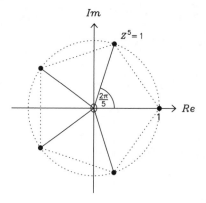

(b) $\quad z^5 = 1 + i = \sqrt{2}\, e^{i(\pi/4 + 2\pi n)}$ where $n = 0, \pm 1, \pm 2, \pm 3, \ldots$

$$\therefore \quad z = 2^{1/10}\, e^{i(\pi/4 + 2\pi n)/5} = \underline{2^{1/10}\, e^{i\pi(1 + 8n)/20}} \qquad \text{for} \quad n = 0, 1, 2, 3, 4$$

(c) $\quad (z+1)^5 = 1 = e^{i2\pi n}$ for $\pm$ integer $n$

$$\therefore \quad z + 1 = e^{i2\pi n/5} \Rightarrow \underline{z = e^{i2\pi n/5} - 1} \qquad \text{for} \quad n = 0, 1, 2, 3, 4$$

(d) $\quad \left(\dfrac{z+1}{z}\right)^5 = 1 = e^{i2\pi n} \quad$ for $\pm$ integer $n$

$\quad \therefore \ z+1 = z\, e^{i2\pi n/5}$

$\quad \therefore \ 1 = z\left(e^{i2\pi n/5} - 1\right) \quad \Rightarrow \quad \underline{z = \dfrac{1}{e^{i2\pi n/5} - 1}} \quad$ for $n = 1,\, 2,\, 3,\, 4$

This is an interesting case because a seemingly $5^{\text{th}}$-order polynomial has only 4 roots: $n=0$ is not a permissible solution because the denominator, $e^{i2\pi n/5} - 1$, is then zero. The difficulty can also be seen two lines earlier, when $n=0$ yields $z+1 = z$ and leads to the inconsistent conclusion that $1 = 0$. The resolution of the paradox is easily found in that a substitution of the binomial expansion of $(z+1)^5$ into the equation $(z+1)^5 = z^5$ shows that it reduces to a fourth-order polynomial: $5z^4 + 10z^3 + 10z^2 + 5z + 1 = 0$. The real and imaginary part of $z$ can be obtained in the usual way by multiplying the top and bottom of the expression above by the complex conjugate of the denominator.

$$\left(e^{i2\pi n/5} - 1\right)\left(e^{-i2\pi n/5} - 1\right)$$
$$= 2\left[1 - \cos(2\pi n/5)\right]$$

---

**(6)** By considering $e^{iA}$ and $e^{iB}$, derive the compound-angle formulae for $\sin(A+B)$ and $\cos(A+B)$.

---

$$e^{i(A+B)} = e^{iA}\, e^{iB} \quad \text{and} \quad e^{i\theta} = \cos\theta + i\sin\theta$$

$$\therefore \quad \cos(A+B) + i\sin(A+B) = (\cos A + i\sin A)(\cos B + i\sin B)$$

$$= \cos A \cos B - \sin A \sin B$$

$$+ i(\sin A \cos B + \cos A \sin B)$$

Equating real parts $\Rightarrow \quad \underline{\cos(A+B) = \cos A \cos B - \sin A \sin B}$

Equating imaginary parts $\Rightarrow \quad \underline{\sin(A+B) = \sin A \cos B + \cos A \sin B}$

---

**(7)** Use de Moivre's theorem to find expansions of $\sin 4\theta$ and $\cos 4\theta$ in terms of powers of $\sin\theta$ and $\cos\theta$.

---

$$\cos 4\theta + i\sin 4\theta = (\cos\theta + i\sin\theta)^4$$

$$= \cos^4\theta + 4i\cos^3\theta \sin\theta + 6i^2\cos^2\theta \sin^2\theta + 4i^3\cos\theta \sin^3\theta + i^4\sin^4\theta$$

$$= \cos^4\theta - 6\cos^2\theta \sin^2\theta + \sin^4\theta + i\left[4\cos^3\theta \sin\theta - 4\cos\theta \sin^3\theta\right]$$

Equating real parts $\Rightarrow$ $\underline{\cos 4\theta = \cos^4\theta - 6\cos^2\theta\sin^2\theta + \sin^4\theta}$

Equating imaginary parts $\Rightarrow$ $\underline{\sin 4\theta = 4\cos^3\theta\sin\theta - 4\cos\theta\sin^3\theta}$

This analysis with de Moivre's theorem is much more straightforward than the alternative procedure of repeatedly using the double-angle formulae for $\sin 2\theta$ and $\cos 2\theta$.

---

**(8)** Show that $\cos^6\theta = (\cos 6\theta + 6\cos 4\theta + 15\cos 2\theta + 10)/32$.

---

$$\cos^6\theta = \left(\frac{e^{i\theta} + e^{-i\theta}}{2}\right)^6$$

$$= \frac{e^{i6\theta} + 6e^{i4\theta} + 15e^{i2\theta} + 20 + 15e^{-i2\theta} + 6e^{-i4\theta} + e^{-i6\theta}}{2^6}$$

$$= \frac{1}{32}\left[\frac{(e^{i6\theta} + e^{-i6\theta})}{2} + 6\frac{(e^{i4\theta} + e^{-i4\theta})}{2} + 15\frac{(e^{i2\theta} + e^{-i2\theta})}{2} + \frac{20}{2}\right]$$

$$= \underline{\tfrac{1}{32}(\cos 6\theta + 6\cos 4\theta + 15\cos 2\theta + 10)}$$

---

**(9)** Use the definitions of the hyperbolic functions to show that
$$\sinh(x+y) = \sinh x \cosh y + \cosh x \sinh y$$
and derive a similar expansion for $\cosh(x+y)$.

---

$$\sinh\theta = \frac{e^\theta - e^{-\theta}}{2} \quad \text{and} \quad \cosh\theta = \frac{e^\theta + e^{-\theta}}{2}$$

$$\therefore \quad \sinh x \cosh y = \left(\frac{e^x - e^{-x}}{2}\right)\left(\frac{e^y + e^{-y}}{2}\right) = \frac{e^{x+y} + e^{x-y} - e^{-x+y} - e^{-x-y}}{4}$$

$$\cosh x \sinh y = \left(\frac{e^x + e^{-x}}{2}\right)\left(\frac{e^y - e^{-y}}{2}\right) = \frac{e^{x+y} - e^{x-y} + e^{-x+y} - e^{-x-y}}{4}$$

$$\therefore \quad \underline{\sinh x \cosh y + \cosh x \sinh y} = \frac{e^{x+y} - e^{-(x+y)}}{2}$$

$$= \underline{\sinh(x+y)}$$

$$\cosh x \cosh y = \left(\frac{e^x+e^{-x}}{2}\right)\left(\frac{e^y+e^{-y}}{2}\right) = \frac{e^{x+y}+e^{x-y}+e^{-x+y}+e^{-x-y}}{4}$$

$$\sinh x \sinh y = \left(\frac{e^x-e^{-x}}{2}\right)\left(\frac{e^y-e^{-y}}{2}\right) = \frac{e^{x+y}-e^{x-y}-e^{-x+y}+e^{-x-y}}{4}$$

$$\therefore \ \underline{\cosh x \cosh y + \sinh x \sinh y} = \frac{e^{x+y}+e^{-(x+y)}}{2}$$

$$= \underline{\cosh(x+y)}$$

---

**(10)** Find:

$$\text{(a)} \ \sum_{k=0}^{\infty}\frac{\cos(k\theta)}{k!} \quad \text{and} \quad \text{(b)} \ \int e^{ax}\sin(bx)\,dx$$

where the constants $a$ and $b$ are real.

---

(a) $$\sum_{k=0}^{\infty}\frac{\cos(k\theta)}{k!} = \sum_{k=0}^{\infty}\frac{\mathcal{Re}\{e^{ik\theta}\}}{k!}$$

$$= \sum_{k=0}^{\infty}\mathcal{Re}\left\{\frac{e^{ik\theta}}{k!}\right\} = \mathcal{Re}\left\{\sum_{k=0}^{\infty}\frac{(e^{i\theta})^k}{k!}\right\}$$

But $$\sum_{k=0}^{\infty}\frac{\Phi^k}{k!} = 1+\Phi+\frac{\Phi^2}{2!}+\frac{\Phi^3}{3!}+\frac{\Phi^4}{4!}+\cdots = \exp(\Phi)$$

$$\therefore \ \sum_{k=0}^{\infty}\frac{(e^{i\theta})^k}{k!} = \exp(e^{i\theta})$$

$$= \exp(\cos\theta+i\sin\theta) = e^{\cos\theta}\,e^{i\sin\theta}$$

i.e. $$\sum_{k=0}^{\infty}\frac{\cos(k\theta)}{k!} = \mathcal{Re}\left\{e^{\cos\theta}[\cos(\sin\theta)+i\sin(\sin\theta)]\right\}$$

$$= \underline{e^{\cos\theta}\cos(\sin\theta)}$$

The corresponding sum of $\sin(k\theta)/k!$, from $k=0$ to $\infty$, can be written straight down from the penultimate line, as being $e^{\cos\theta}\sin(\sin\theta)$, because the analysis is identical apart from the replacement of $\mathcal{Re}\{\}$ with $\mathcal{Im}\{\}$.

(b)    $e^{ax}\sin(bx) = e^{ax}\mathcal{I}m\{e^{ibx}\} = \mathcal{I}m\{e^{ax}e^{ibx}\}$

$$\therefore \int e^{ax}\sin(bx)\,dx = \int \mathcal{I}m\{e^{x(a+ib)}\}\,dx$$

$$= \mathcal{I}m\left\{\int e^{x(a+ib)}\,dx\right\} = \mathcal{I}m\left\{\frac{e^{x(a+ib)}}{a+ib} + C\right\}$$

But    $\dfrac{e^{x(a+ib)}}{a+ib} = \dfrac{e^{ax}e^{ibx}}{(a+ib)} \times \dfrac{(a-ib)}{(a-ib)} = \dfrac{e^{ax}(a-ib)e^{ibx}}{a^2+b^2}$

$$\therefore \int e^{ax}\sin(bx)\,dx = \mathcal{I}m\left\{\frac{e^{ax}}{a^2+b^2}(a-ib)\big[\cos(bx) + i\sin(bx)\big] + C\right\}$$

$K = \mathcal{I}m\{C\}$

$$= \frac{e^{ax}}{a^2+b^2}\big[a\sin(bx) - b\cos(bx)\big] + K$$

We can easily write down the integral of $e^{ax}\cos(bx)$ from the penultimate line by replacing $\mathcal{I}m\{\}$ with $\mathcal{R}e\{\}$, as with the summation in part (a), so we have really done two integrals for the price of one! We could have evaluated this integral by using integration by parts twice, but the complex number formulation accomplishes the same task with just one simple exponential integration. We should also note that the constants $a$ and $b$ were implicitly assumed to be real (as must be $K$).

# 8 Vectors

## 8.1 Definition and Cartesian coordinates

The simplest way to visualize a *vector* is to hold a pencil up in the air. The length of the pencil is a *scalar* quantity, a single number with an associated *unit*, 0.15 metres, say, for a typical pencil. By describing the pencil by a vector, we can give extra information: we can convey not only its length but also where it is pointing. Vectors are therefore defined by both a *magnitude* (or *modulus*), a positive number, normally with an associated unit, and a *direction* in space. Many physical quantities, such as position and velocity, are vectors which must be manipulated in slightly different ways to scalars like mass, length, and time.

We will use bold script, as in **a**, to identify vectors. Other conventions, that are better suited to handwritten text, include underlining, $\underline{a}$ or $\underset{\sim}{a}$, and an over-placed arrow, $\overrightarrow{a}$ or $\overset{\rightharpoonup}{a}$. The modulus, or length, of the vector is denoted by $|\mathbf{a}|$. Those with a magnitude of 1 are called *unit vectors* and are often indicated by an over-placed 'hat', as in $\widehat{\mathbf{a}}$.

$$\widehat{\mathbf{a}} = \frac{\mathbf{a}}{|\mathbf{a}|}$$

Unlike scalars, more than one number is needed to describe a vector. Let us start thinking about this by confining our pencil to lie in a plane, for example by dropping it onto a two-dimensional surface like a table. Only two numbers are then required to define its length and orientation, namely the displacements in the $x$ and $y$ directions required to get from one end of the pencil to the other. These offsets are called the *components* of the vector, and are scalar quantities that tell us how far to move along unit vectors parallel with the $x$ and $y$ axes. That is, placing the *origin* of our system at the blunt end of the pencil, we can express the *position vector* of the sharp end by $x\,\mathbf{i} + y\,\mathbf{j}$ where we have introduced perpendicular unit vectors **i** and **j** which point along the $x$ and $y$ axes respectively.

Now lift the sharp end of the pencil off the table, leaving the blunt end at the origin. To describe its length and orientation we need a third component in the direction of the $z$-axis, and another unit vector **k** at right-angles to both **i** and **j**. Mathematically, we say that we have moved from a two-dimensional to a three-dimensional *vector space*. The position vector of the sharp end is now $x\,\mathbf{i} + y\,\mathbf{j} + z\,\mathbf{k}$, which is written in shorthand as $(x, y, z)$. Denoting this vector by **r**, Pythagoras' theorem gives its modulus-squared as

$$|\mathbf{r}|^2 = x^2 + y^2 + z^2 \tag{8.1}$$

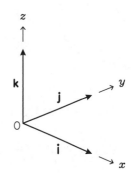

A *suffix notation* can also be used to describe all three components of a vector in the compact form $x_m$ with $m$ free to take the values 1, 2, and 3, so that $x_1 = x$, $x_2 = y$, and $x_3 = z$.

We call the mutually perpendicular unit vectors **i**, **j**, and **k** the *basis* of the *Cartesian* coordinate system. By convention, the **k** vector points upwards at us if we imagine ourselves peering down onto a standard $x$-$y$ plane, or graph paper. It's helpful to train the right-hand to ape these **i**, **j**, and **k** axes using the thumb, first, and second fingers respectively.

Although we will confine ourselves to three-dimensional problems, it is worth mentioning that many, but not all, of the ways in which we will learn to manipulate vectors are applicable to N-dimensional spaces where N could be any positive integer. Indeed, as a result of the power of modern computers, scientists and engineers commonly manipulate vectors with $N = 10^6$; the components are saved in memory as a set of N scalar values $(a_1, a_2, \ldots, a_N)$.

## 8.2  Addition, subtraction, and scalar multiplication

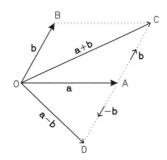

As well as having a modulus and direction, vectors obey some simple algebraic rules. The *addition* of vectors **a** and **b** to form a new vector **c** is written as **c** = **a** + **b**. Since two vectors are equal only if all of their components are equal, $c_m = a_m + b_m$ for all possible values of $m$; which, in a three-dimensional space, is a shorthand way of expressing three equations.

Subtraction of **b** from **a** is achieved by reversing the direction of **b** and adding: **d** = **a** + (−**b**), so that $d_m = a_m - b_m$. In the triangle formed by the origin and the points A and B, with position vectors **a** and **b**, the vector **d** = **a** − **b** is the one that joins B to A; this is sometimes written as $\overrightarrow{BA}$.

Multiplication of vectors by scalars is also straightforward: if **b** = $\lambda$**a** then the direction of **b** is the same as that of **a** if $\lambda$ is positive, or opposite to **a** if it is negative; in either case, the modulus of **b** is $|\lambda|$-times that of **a**. An important example concerns the position vector of the mid-point of $\overrightarrow{AB}$, namely

$$\mathbf{e} = \mathbf{a} + \tfrac{1}{2}\mathbf{d} = \mathbf{a} + \tfrac{1}{2}(\mathbf{b} - \mathbf{a}) = \tfrac{1}{2}(\mathbf{a} + \mathbf{b})$$

Armed with these basic rules we can begin to use the general position vector **r** to describe the *loci* of points on lines, curves, and surfaces within a three-dimensional space. Our first example is

$$\mathbf{r} = \mathbf{a} + \lambda(\mathbf{b} - \mathbf{a}) \qquad (8.2)$$

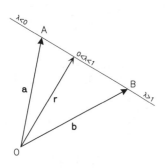

$$\mathbf{r} = (1, 2, 3) + \lambda(3, 2, 1)$$

$$\lambda = \frac{x-1}{3} = \frac{y-2}{2} = \frac{z-3}{1}$$

which is the *vector equation* of the straight line which passes through points A and B: the first term takes us from the origin to a specific point (A) on the line, and the second term takes us an arbitrary distance ($\lambda$) along the direction fixed by the vector joining A to B. If **a** and **b** are the known position vectors of A and B, then $\lambda$ can be found in terms of each component, $(x, y, z)$, of **r**; the *Cartesian equation* of the line is the result of equating the three expressions for $\lambda$.

Our second example is

$$\mathbf{r} = \mathbf{a} + \mu(\mathbf{b} - \mathbf{a}) + \nu(\mathbf{c} - \mathbf{a}) \tag{8.3}$$

which is the vector equation of a plane containing the points A, B, and C. The first term takes us from the origin to point A on the plane, and the second and third allow us to move arbitrary distances (via $\mu$ and $\nu$) along the two directions fixed by the vectors joining A to B and A to C. Provided the three points (A, B, and C) do not lie along a single line, that is they are not *colinear*, the vectors $\mathbf{b} - \mathbf{a}$ and $\mathbf{c} - \mathbf{a}$ are said to *span* the two-dimensional space defined by the plane. This means that by choosing suitable values of $\mu$ and $\nu$ we can get to anywhere on the plane.

Our final example is

$$|\mathbf{r} - \mathbf{a}| = R \tag{8.4}$$

which describes a sphere of radius $R$ centred at the point A: to visualize this note that $\mathbf{r} - \mathbf{a}$ joins A to the general point, and that the modulus sign requires that the length of this vector is always equal to the fixed value $R$.

## 8.3 Scalar product

The *scalar* or *dot product* is one way of 'multiplying vectors together'. It is defined by

$$\mathbf{a} \cdot \mathbf{b} = |\mathbf{a}| |\mathbf{b}| \cos \theta \tag{8.5}$$

where $\theta$ is the angle between the vectors $\mathbf{a}$ and $\mathbf{b}$. Since $\cos(-\theta) = \cos(\theta)$ there is no need for a convention to determine 'which way round' to measure the angle, and the result is a simple scalar. If $\mathbf{b}$ is taken to be a unit vector, then the geometric meaning of the dot product becomes clear: it is the length $|\mathbf{a}| \cos \theta$ which is the *projection* of $\mathbf{a}$ in the direction of $\mathbf{b}$. If we take $\mathbf{a}$ to be unit instead, then $\mathbf{a} \cdot \mathbf{b}$ can be viewed as the projection of $\mathbf{b}$ along the direction of $\mathbf{a}$. Thus dot products play an important rôle when vectors, such as forces, need to be resolved along particular directions.

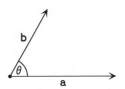

If $\theta = \pi/2$ then the dot product is zero, and the vectors $\mathbf{a}$ and $\mathbf{b}$ are said to be *orthogonal*. The sign of the dot product is positive if $\theta$ is acute and negative if it's obtuse. If $\theta = 0$, the vectors $\mathbf{a}$ and $\mathbf{b}$ are parallel and $\mathbf{a} \cdot \mathbf{b} = |\mathbf{a}| |\mathbf{b}|$. In particular $\mathbf{a} \cdot \mathbf{a} = |\mathbf{a}|^2$, which gives a useful way of finding the modulus-squared of a vector. Also

$$|\mathbf{b} - \mathbf{c}|^2 = (\mathbf{b} - \mathbf{c}) \cdot (\mathbf{b} - \mathbf{c}) = |\mathbf{b}|^2 + |\mathbf{c}|^2 - 2\mathbf{b} \cdot \mathbf{c} \tag{8.6}$$

because we are allowed to 'multiply the brackets out' just as in normal algebra. In fact, eqn (8.6) provides a remarkably compact proof of the cosine rule for triangles that we introduced in eqn (3.23).

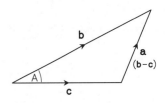

We next need to find a way of evaluating the dot product in the Cartesian system. It is given by

$$\mathbf{a} \cdot \mathbf{b} = (a_1\mathbf{i} + a_2\mathbf{j} + a_3\mathbf{k}) \cdot (b_1\mathbf{i} + b_2\mathbf{j} + b_3\mathbf{k}) = a_1 b_1 + a_2 b_2 + a_3 b_3 \tag{8.7}$$

where we have again multiplied out, and noted that $\mathbf{i} \cdot \mathbf{j} = \mathbf{j} \cdot \mathbf{k} = \mathbf{k} \cdot \mathbf{i} = 0$ and that, being unit vectors, $|\mathbf{i}|^2 = |\mathbf{j}|^2 = |\mathbf{k}|^2 = 1$. Hence the dot product is calculated by summing the products of corresponding components.

Taking the dot product of a vector equation is useful because it produces a scalar equation which can be manipulated according to all the normal algebraic rules. Division by a vector is not defined, however, so never do it!

## 8.4 Vector product

Another way of multiplying together $\mathbf{a}$ and $\mathbf{b}$ is with a *vector*, or *cross*, product whose result is also a vector

$$\mathbf{a} \times \mathbf{b} = |\mathbf{a}| \, |\mathbf{b}| \sin \theta \, \hat{\mathbf{u}} \qquad (8.8)$$

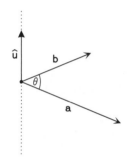

The magnitude of the cross product is equal to $|\mathbf{a}| \, |\mathbf{b}| \sin \theta$, where $\theta$ is again the angle between $\mathbf{a}$ and $\mathbf{b}$, and its direction is specified by the unit vector $\hat{\mathbf{u}}$. The orientation of $\hat{\mathbf{u}}$ is perpendicular to both $\mathbf{a}$ and $\mathbf{b}$, and hence *normal* to the plane containing them, being uniquely defined by the 'right-hand screw rule'. That is to say, if the right hand is held as though it were gripping a screw-driver, and the curl of the fingers is made to indicate the sense of rotation needed to go from $\mathbf{a}$ to $\mathbf{b}$, then the out-stretched thumb points along $\hat{\mathbf{u}}$. This prescription immediately tells us that the direction of $\mathbf{a} \times \mathbf{b}$ is opposite to that of $\mathbf{b} \times \mathbf{a}$

$$\mathbf{a} \times \mathbf{b} = -\mathbf{b} \times \mathbf{a} \qquad (8.9)$$

because the sense of rotation, and therefore the orientation of the hand (and thumb), has to be reversed to go from $\mathbf{b}$ to $\mathbf{a}$. Also, since $\sin \theta = 0$ if $\theta = 0$, the cross product will be zero if two vectors are parallel to each other.

To evaluate the vector product of $\mathbf{a}$ and $\mathbf{b}$ in terms of their Cartesian components, we first need to note the results of taking cross products between their basis vectors: $\mathbf{i} \times \mathbf{i} = \mathbf{j} \times \mathbf{j} = \mathbf{k} \times \mathbf{k} = 0$, and $\mathbf{i} \times \mathbf{j} = \mathbf{k}$, $\mathbf{j} \times \mathbf{i} = -\mathbf{k}$, and so on. The expansion of $(a_1 \mathbf{i} + a_2 \mathbf{j} + a_3 \mathbf{k}) \times (b_1 \mathbf{i} + b_2 \mathbf{j} + b_3 \mathbf{k})$ then leads to

$$\mathbf{a} \times \mathbf{b} = (a_2 b_3 - b_2 a_3) \mathbf{i} + (a_3 b_1 - b_3 a_1) \mathbf{j} + (a_1 b_2 - b_1 a_2) \mathbf{k} \qquad (8.10)$$

and we'll see a succinct way of remembering this in section 9.3.

Geometrically, the magnitude of a cross product is the area of a parallelogram, or twice the area of the triangle, whose adjacent sides are given by $\mathbf{a}$ and $\mathbf{b}$. Indeed, by definition, the *vector area* of the parallelogram is given by $\mathbf{a} \times \mathbf{b}$, with the direction giving a unique normal to the planar surface.

Area of triangle $= \frac{1}{2} |\mathbf{a} \times \mathbf{b}|$

To see the cross product in action we can post-multiply both sides of eqn (8.2) by $(\mathbf{b} - \mathbf{a})$ and, noting that $(\mathbf{b} - \mathbf{a}) \times (\mathbf{b} - \mathbf{a}) = 0$, eliminate the scalar parameter $\lambda$ to obtain

$$\mathbf{r} \times (\mathbf{b} - \mathbf{a}) = \mathbf{a} \times (\mathbf{b} - \mathbf{a}) = \mathbf{a} \times \mathbf{b} \qquad (8.11)$$

which is an alternative form of the vector equation of a line.

## 8.5 Scalar triple product

We next explore the concept of 'multiplying three vectors', **a**, **b**, and **c**. The easiest way of doing this is the *scalar triple product* which is written **a**.(**b** × **c**), or sometimes $[\mathbf{a}, \mathbf{b}, \mathbf{c}]$, and produces a scalar. In component form this gives

$$\mathbf{a} \cdot (\mathbf{b} \times \mathbf{c}) = a_1(b_2 c_3 - b_3 c_2) + a_2(b_3 c_1 - b_1 c_3) + a_3(b_1 c_2 - b_2 c_1) \quad (8.12)$$

where again a more compact notation will be introduced in section 9.3.

The scalar triple product measures the volume of the *parallelepiped* whose edges are formed by the vectors **a**, **b**, and **c**. We saw in section 8.4 that **b** × **c** gives the area of the parallelogram with sides **b** and **c**. The dot product with **a** multiplies this area by the perpendicular height, given by the projection of **a** along the normal to the base parallelogram, and thus generates the volume. Since volume is a fixed physical quantity, the magnitude of the scalar triple product is independent of the order of **a**, **b**, and **c**, but we must take some care with its sign. Specifically, the *cyclic combinations* of **a**, **b**, and **c** are all equal:

$$\mathbf{a} \cdot (\mathbf{b} \times \mathbf{c}) = \mathbf{b} \cdot (\mathbf{c} \times \mathbf{a}) = \mathbf{c} \cdot (\mathbf{a} \times \mathbf{b})$$

The *anti-cyclic* ones, $\mathbf{a} \cdot (\mathbf{c} \times \mathbf{b}) = \mathbf{b} \cdot (\mathbf{a} \times \mathbf{c}) = \mathbf{c} \cdot (\mathbf{b} \times \mathbf{a})$, are all equal to $-\mathbf{a} \cdot (\mathbf{b} \times \mathbf{c})$, as can be seen most clearly using eqn (8.9).

If any two vectors in a scalar triple product are parallel, or the same, then its value must be zero because it is the volume of a parallelepiped of zero height. This provides a convenient test of whether three vectors, **a**, **b**, and **c**, lie in a plane: if so, the scalar triple product **a**.(**b** × **c**) is necessarily zero. Three non-*coplanar* vectors are said to span a three-dimensional vector space in the sense that any point in it can be expressed in the form

$$\mathbf{r} = l\mathbf{a} + m\mathbf{b} + n\mathbf{c} \quad (8.13)$$

where $l$, $m$, and $n$ are scalars. If two of the three vectors are parallel, then the vectors only span a two-dimensional space and are said to be *linearly dependent* in that we can always find scalars $p$ and $q$ such that

$$\mathbf{c} = p\mathbf{a} + q\mathbf{b} \quad (8.14)$$

To find $l$ in eqn (8.13), for example, we take the scalar product of both sides with **b** × **c**, generating a scalar equation, and then divide both sides by **a**.(**b** × **c**).

To end this section let us use the scalar triple product to derive an alternative form for the vector equation of a plane given by eqn (8.3). First, take the scalar product of both sides of this equation with the vector

$$\mathbf{n} = (\mathbf{b} - \mathbf{a}) \times (\mathbf{c} - \mathbf{a}) = \mathbf{b} \times \mathbf{c} + \mathbf{a} \times \mathbf{b} + \mathbf{c} \times \mathbf{a}$$

noting it makes no odds whether we pre- or post-multiply. We choose this form of **n** to ensure that both of the arbitrary constants, $\mu$ and $\nu$, are eliminated: this occurs because they are multiplied by scalar triple products which, because they

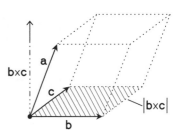

Volume of tetrahedron
$$= \tfrac{1}{6} |\mathbf{a} \cdot (\mathbf{b} \times \mathbf{c})|$$

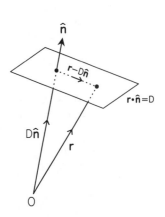

contain two parallel vectors, have values of zero. The direction of this vector is normal to the plane containing the vectors $(\mathbf{b} - \mathbf{a})$ and $(\mathbf{c} - \mathbf{a})$. We obtain

$$\frac{\mathbf{r} \cdot \mathbf{n}}{|\mathbf{n}|} = \frac{\mathbf{a} \cdot (\mathbf{b} \times \mathbf{c})}{|\mathbf{n}|} = D \tag{8.15}$$

where we have divided both sides by $|\mathbf{n}|$ to make the left-hand side the scalar product of $\mathbf{r}$ with a unit vector, $\hat{\mathbf{n}}$. This provides a clear picture of the geometry of a plane: the left-hand side of eqn (8.15) is the projection of $\mathbf{r}$ along the unit normal to the plane, so that $D$ is the perpendicular distance from the origin to the plane. The Cartesian equation of the plane can be obtained by substituting $\mathbf{r} = (x, y, z)$ into eqn (8.15).

$$\mathbf{r} \cdot (1, 2, 3) = 5$$

$$\therefore \ D = \frac{5}{|(1,2,3)|} = \frac{5}{\sqrt{14}}$$

$$x + 2y + 3z = 5$$

## 8.6 Vector triple product

Another way of combining three vectors is with the *vector triple product*, written $\mathbf{a} \times (\mathbf{b} \times \mathbf{c})$, which produces a vector. It can be simplified using

$$\mathbf{a} \times (\mathbf{b} \times \mathbf{c}) = \underbrace{(\mathbf{a} \cdot \mathbf{c})\mathbf{b}}_{AB} - \underbrace{(\mathbf{a} \cdot \mathbf{b})\mathbf{c}}_{AC} \quad\quad \tag{8.16}$$

which can be remembered as the 'ABACAB' identity. An example of the use of eqn (8.16) is given by one method of finding the line of intersection of two planes. Following eqn (8.15) we can write these two planes as

$$\mathbf{r} \cdot \mathbf{a} = u \quad \text{and} \quad \mathbf{r} \cdot \mathbf{b} = v$$

If we now consider the vector triple product $\mathbf{r} \times (\mathbf{a} \times \mathbf{b})$, and apply the ABACAB identity, we obtain

$$\mathbf{r} \times (\mathbf{a} \times \mathbf{b}) = (\mathbf{r} \cdot \mathbf{b})\mathbf{a} - (\mathbf{r} \cdot \mathbf{a})\mathbf{b} = v\mathbf{a} - u\mathbf{b}$$

where we have used the fact that $\mathbf{r}$ must lie on both planes. A comparison with eqn (8.11) shows that we have effortlessly recovered the vector equation of the line of intersection, the direction being given by $\mathbf{a} \times \mathbf{b}$.

## 8.7 Polar coordinates

$$x = r \cos \theta$$
$$y = r \sin \theta$$

$$r^2 = x^2 + y^2$$
$$\theta = \tan^{-1}(y/x)$$

The location of a point on a graph, or in two-dimensional space, is usually specified with Cartesian coordinates $(x, y)$. In section 7.3, however, we noted that an alternative way of defining the position was in terms of the distance from the origin, $r$, and the anticlockwise angle, $\theta$, made by this 'radius' with the positive $x$-axis; the latter are known as *polar* coordinates, and written as $(r, \theta)$. The relationship between the Cartesian and polar forms is given in eqns (7.9) and (7.10).

When working in three dimensions, there are two commonly used generalizations of the polar formulation. The first is very easy: $x$ and $y$ are replaced with $r$

and $\theta$, as above, but $z$ is kept unchanged. Thus the triplet $(r, \theta, z)$ constitutes *cylindrical polar* coordinates. In the second approach, the location of a point is specified by its distance from the origin, or radius $r$, and two angles, $\theta$ and $\phi$, which correspond to its co-latitude and longitude. That is to say, $\theta$ is measured from the $z$-axis so that $0°$ is 'due north', $90°$ is in the $x$-$y$ plane (with $z = 0$) and $180°$ is in the direction of the 'south pole'; the longitude $\phi$ is the anticlockwise angle from the positive $x$-axis to the projection of the radius on to the 'equatorial' plane. Following the global overtones, $(r, \theta, \phi)$ are called *spherical polar* coordinates. A bit of thought, and elementary trigonometry, shows that $r$, $\theta$, and $\phi$ are related to their Cartesian counterparts by

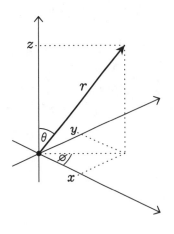

$$x = r \sin\theta \cos\phi$$
$$y = r \sin\theta \sin\phi \qquad (8.17)$$
$$z = r \cos\theta$$

The reverse transformations can also be ascertained, of which the most useful is the familiar result that $r^2 = x^2 + y^2 + z^2$.

Although it may seem unnecessary to have different coordinate systems when a Cartesian choice will suffice, and is so much easier to manipulate, we have already seen the value of the polar formulation in chapter 7. The advantages of working in a reference system which matches the natural geometry of the problem at hand, such as a simplification of the related algebra, will also be met in chapter 11. While we have only considered the cases relevant for rectangular, cylindrical, and spherical situations here, there are many alternatives; in practice, they are rarely used.

## 8.8 Worked examples

> **(1)** State which of the following quantities can be described by vectors:
> (a) temperature, (b) magnetic field, (c) acceleration, (d) force,
> (e) molecular weight, (f) area.

(a) Temperature is always a <u>scalar</u> quantity because it makes no sense to talk about the direction of a temperature; it can vary with position, however, making its *gradient* a vector. A metal cube heated and cooled at its opposite corners develops a temperature distribution, $T(\mathbf{r})$, a *scalar function* of the position vector $\mathbf{r}$.

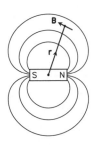

(b) Magnetic field is a <u>vector</u> quantity. A compass can be used to give its direction at any point. In general it is a *vector function* of position, written as $\mathbf{B}(\mathbf{r})$, since, as illustrated by the field of a bar magnet, both the direction and magnitude of the field are usually functions of the position vector $\mathbf{r}$.

$$\mathbf{a} = \frac{d\mathbf{v}}{dt} = \frac{d}{dt}\left(\frac{d\mathbf{r}}{dt}\right) = \frac{d^2\mathbf{r}}{dt^2}$$

$$\mathbf{F} = \frac{d}{dt}(m\mathbf{v}) = m\underbrace{\frac{d\mathbf{v}}{dt}}_{\mathbf{a}} + \mathbf{v}\underbrace{\frac{dm}{dt}}_{0}$$

(c) & (d) Acceleration is the rate of change of velocity with time, and velocity is the rate of change of position with time. So, since position is given by the vector $\mathbf{r}$, both velocity $\mathbf{v} = d\mathbf{r}/dt$ and acceleration $\mathbf{a} = d\mathbf{v}/dt$ must also be <u>vector</u> quantities. Newton's *Second Law of motion* states that force $\mathbf{F}$ is equal to the rate of change of momentum with time, and momentum is defined as the product of (the scalar) mass $m$ and velocity (vector) $\mathbf{v}$. If the mass of a body remains constant, then the Second Law reduces to the '$F = ma$' relation familiar from school physics, but with both force $\mathbf{F}$ and acceleration $\mathbf{a}$ as vector quantities.

(e) Molecular weight is a <u>scalar</u> quantity because it is the sum of the atomic weights of the atoms which comprise a molecule, and these weights are defined to be the mass of each atom relative to $\frac{1}{12}$ of the (scalar) mass of a $^{12}_{6}C$ atom. In other areas of science 'weight' is reserved for vector quantities. For example, the weight of a woman of mass $m$ is the force she exerts on a set of bathroom scales; by Newton's Second Law this is given by $m\mathbf{g}$, where $\mathbf{g}$ is the vector acceleration due to gravity, which is directed towards the Earth's centre.

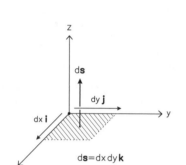

(f) Somewhat surprisingly, area can be considered as a <u>vector</u> quantity. For a rectangular table with sides $a$ and $b$, for example, it is defined to have a magnitude $ab$ and a direction normal to the surface given by the 'right-hand screw rule' (when rotating from $a$ to $b$). Although most surfaces are far more complicated, any vanishingly small portion can be treated as a plane and the total surface area obtained by the vector addition of all such segments of area that comprise the whole $\left(\mathbf{s} = \int d\mathbf{s}\right)$.

---

**(2)** The four points A, B, C, and D respectively have position vectors

$$\mathbf{a} = (1,2,3), \quad \mathbf{b} = (2,0,1), \quad \mathbf{c} = (1,1,1), \quad \text{and} \quad \mathbf{d} = (5,2,5)$$

Calculate $2\mathbf{a} - 3\mathbf{b} - 5\mathbf{c} + \frac{1}{2}\mathbf{d}$, and the mid-points of $\overrightarrow{BC}$ and $\overrightarrow{AD}$.

---

$$2\mathbf{a} - 3\mathbf{b} - 5\mathbf{c} + \tfrac{1}{2}\mathbf{d} = \left(\, 2\times1 - 3\times2 - 5\times1 + \tfrac{1}{2}\times5,\right.$$
$$2\times2 - 3\times0 - 5\times1 + \tfrac{1}{2}\times2,$$
$$\left. 2\times3 - 3\times1 - 5\times1 + \tfrac{1}{2}\times5 \,\right)$$
$$= \left(-\tfrac{13}{2}, 0, \tfrac{1}{2}\right)$$

Mid-point of $\overrightarrow{BC} = \mathbf{b} + \tfrac{1}{2}(\mathbf{c} - \mathbf{b}) = \tfrac{1}{2}(\mathbf{b} + \mathbf{c}) = \tfrac{1}{2}(3,1,2)$

Mid-point of $\overrightarrow{AD} = \tfrac{1}{2}(\mathbf{a} + \mathbf{d}) = \tfrac{1}{2}(6,4,8) = (3,2,4)$

**(3)** Taking **a**, **b**, **c**, and **d** from exercise (2), find: (a) the vector equation of the line passing through A and C; (b) the vector equation of the line passing through the mid-points of $\overrightarrow{AB}$ and $\overrightarrow{CD}$; (c) the Cartesian equations of the lines defined by $\mathbf{r} = \mathbf{a} + \lambda\mathbf{b}$ and $\mathbf{r} = \mathbf{c} + \mu\mathbf{d}$.

(a)  We need to get from the origin to a point on the line, say A, and then have the freedom to move an arbitrary scalar distance, given by the value of $\lambda$, along the direction $\overrightarrow{AC} = \mathbf{c} - \mathbf{a}$. This gives

$$\mathbf{r} = \mathbf{a} + \lambda(\mathbf{c} - \mathbf{a}) = (1, 2, 3) + \lambda(1-1, 1-2, 1-3)$$

$$= \underline{(1, 2, 3) - \lambda(0, 1, 2)}$$

(b)   Mid-point of $\overrightarrow{AB}$, $\mathbf{e} = \frac{1}{2}(\mathbf{a} + \mathbf{b}) = \frac{1}{2}(3, 2, 4)$

Mid-point of $\overrightarrow{CD}$, $\mathbf{f} = \frac{1}{2}(\mathbf{c} + \mathbf{d}) = \frac{1}{2}(6, 3, 6)$

$\therefore$ Equation of line is $\mathbf{r} = \mathbf{e} + \lambda(\mathbf{f} - \mathbf{e}) = \underline{\frac{1}{2}(3, 2, 4) + \frac{\lambda}{2}(3, 1, 2)}$

(c)    $\mathbf{r} = \mathbf{a} + \lambda\mathbf{b} \Rightarrow (x, y, z) = (1 + 2\lambda, 2, 3 + \lambda)$

$$\therefore \lambda = \underline{\frac{x-1}{2} = z - 3 \quad \text{and} \quad y = 2}$$

$\mathbf{r} = \mathbf{c} + \lambda\mathbf{d} \Rightarrow (x, y, z) = (1 + 5\lambda, 1 + 2\lambda, 1 + 5\lambda)$

$$\therefore \lambda = \underline{\frac{x-1}{5} = \frac{y-1}{2} = \frac{z-1}{5}}$$

**(4)** Taking **a**, **b**, **c**, and **d** from exercise (2), find the angle between **c** and **d** and evaluate $(\mathbf{a} \cdot \mathbf{c})\mathbf{b}$.

$$\cos(\hat{COD}) = \frac{\mathbf{c} \cdot \mathbf{d}}{|\mathbf{c}||\mathbf{d}|} = \frac{1 \times 5 + 1 \times 2 + 1 \times 5}{\sqrt{1^2 + 1^2 + 1^2}\sqrt{5^2 + 2^2 + 5^2}} = \frac{12}{\sqrt{3} \times 3\sqrt{6}}$$

$$\therefore \hat{COD} = \cos^{-1}\left(\frac{2\sqrt{2}}{3}\right) = \underline{19.5°}$$

$(\mathbf{a} \cdot \mathbf{c})\mathbf{b} = \left[1 \times 1 + 2 \times 1 + 3 \times 1\right](2, 0, 1) = 6(2, 0, 1) = \underline{(12, 0, 6)}$

We should emphasize that terms like $(\mathbf{a} \cdot \mathbf{c})\mathbf{b}$ are vectors, and are best thought of as the scalar '**a** dot **c**' *lots of* the vector **b**.

**(5)** Use the dot product to show $\cos(A \pm B) = \cos A \cos B \mp \sin A \sin B$.

Consider two vectors $\mathbf{v}_A$ and $\mathbf{v}_B$ in the $x\text{-}y$ plane ($z = 0$) which make angles of $A$ and $B$ with the $x$-axis respectively. Then

$$\mathbf{v}_A = |\mathbf{v}_A|(\cos A, \sin A, 0) \quad \text{and} \quad \mathbf{v}_B = |\mathbf{v}_B|(\cos B, \sin B, 0)$$

$$\therefore \quad \mathbf{v}_A \cdot \mathbf{v}_B = |\mathbf{v}_A||\mathbf{v}_B|\cos(A - B)$$

and $\quad \mathbf{v}_A \cdot \mathbf{v}_B = |\mathbf{v}_A||\mathbf{v}_B|(\cos A \cos B + \sin A \sin B + 0)$

i.e. $\quad \underline{\cos(A - B) = \cos A \cos B + \sin A \sin B}$

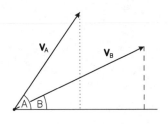

$$\sin(-\theta) = -\sin(\theta)$$

Putting $B = -C \quad \Rightarrow \quad \underline{\cos(A + C) = \cos A \cos C - \sin A \sin C}$

---

**(6)** Taking $\mathbf{a}$, $\mathbf{b}$, $\mathbf{c}$, and $\mathbf{d}$ from exercise (2), use a cross product to find the angle between $\mathbf{c}$ and $\mathbf{d}$. Express the line $\mathbf{r} = \mathbf{a} + \lambda\mathbf{b}$ in the form $\mathbf{r} \times \mathbf{p} = \mathbf{q}$, where $\mathbf{p}$ and $\mathbf{q}$ do not involve $\lambda$.

$$\mathbf{c} \times \mathbf{d} = (1, 1, 1) \times (5, 2, 5)$$

$$= (1 \times 5 - 1 \times 2, \ 1 \times 5 - 1 \times 5, \ 1 \times 2 - 1 \times 5) = (3, 0, -3)$$

$$\sin(\widehat{COD}) = \frac{|\mathbf{c} \times \mathbf{d}|}{|\mathbf{c}||\mathbf{d}|} = \frac{\sqrt{3^2 + 0^2 + 3^2}}{\sqrt{1^2 + 1^2 + 1^2}\sqrt{5^2 + 2^2 + 5^2}} = \frac{3\sqrt{2}}{\sqrt{3} \times 3\sqrt{6}}$$

$$\therefore \quad \widehat{COD} = \sin^{-1}\left(\tfrac{1}{3}\right) = \underline{19.5°} \quad \text{(as before)}$$

Since the cross product of any vector with itself is zero, the line $\mathbf{r} = \mathbf{a} + \lambda\mathbf{b}$ can be expressed in a form that excludes $\lambda$ by taking the cross product of both sides with $\mathbf{b}$.

$$\mathbf{r} \times \mathbf{b} = \mathbf{a} \times \mathbf{b} + \lambda\underbrace{\mathbf{b} \times \mathbf{b}}_{0}$$

But $\quad \mathbf{a} \times \mathbf{b} = (1, 2, 3) \times (2, 0, 1)$

$$= (2 \times 1 - 3 \times 0, \ 3 \times 2 - 1 \times 1, \ 1 \times 0 - 2 \times 2) = (2, 5, -4)$$

$$\therefore \quad \underline{(x, y, z) \times (2, 0, 1) = (2, 5, -4)}$$

---

**(7)** The tetrahedral structure of methane ($CH_4$) can be modelled by placing the four hydrogen atoms at the corners of a cube, with two pairs in a crossed diagonal orientation on opposing faces; the carbon atom is at the centre of the cube. Calculate the tetrahedral bond angle ($H\widehat{C}H$).

Let the carbon atom be at the origin, $(0,0,0)$, and the four hydrogen atoms be located at $(-1,-1,-1)$, $(1,1,-1)$, $(-1,1,1)$ and $(1,-1,1)$. The $\widehat{HCH}$ bond angle can then be ascertained from the dot product of any two of the four vectors representing the displacements from carbon to the various hydrogens. For example,

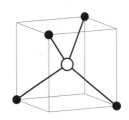

$$\underbrace{(-1,1,1)\bullet(1,-1,1)}_{-1} = \underbrace{|(-1,1,1)|}_{\sqrt{3}}\,\underbrace{|(1,-1,1)|}_{\sqrt{3}}\cos(\widehat{HCH})$$

$$\therefore \text{ Tetrahedral angle, } \widehat{HCH} = \cos^{-1}\left(-\tfrac{1}{3}\right) = \underline{109.5°}$$

---

**(8)** Taking **a**, **b**, **c**, and **d** from exercise (2), find $\mathbf{a}\bullet(\mathbf{b}\times\mathbf{c})$, $\mathbf{a}\bullet(\mathbf{b}\times\mathbf{d})$, and $\mathbf{a}\bullet(\mathbf{c}\times\mathbf{d})$. Which three position vectors are coplanar? Find an equation for this plane in its Cartesian and vector forms, and its perpendicular distance from the origin.

---

$$\mathbf{a}\bullet(\mathbf{b}\times\mathbf{c}) = (1,2,3)\bullet\underbrace{\left[(2,0,1)\times(1,1,1)\right]}_{(-1,-1,2)} = \underline{3}$$

$$\mathbf{a}\bullet(\mathbf{b}\times\mathbf{d}) = (1,2,3)\bullet\underbrace{\left[(2,0,1)\times(5,2,5)\right]}_{(-2,-5,4)} = \underline{0}$$

$$\mathbf{a}\bullet(\mathbf{c}\times\mathbf{d}) = (1,2,3)\bullet\underbrace{\left[(1,1,1)\times(5,2,5)\right]}_{(3,0,-3)} = \underline{-6}$$

If a scalar triple product is zero, then the parallelepiped formed by the three vectors involved has zero volume; this implies that they all lie in a plane. In this case, $\mathbf{a}\bullet(\mathbf{b}\times\mathbf{d}) = 0$ shows that the vectors **a**, **b**, and **d** are coplanar. It follows that any one of them can be written as a linear combination of the other two:

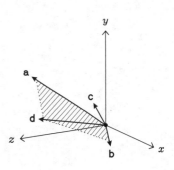

$$(5,2,5) = (1,2,3) + 2(2,0,1) \quad\Longleftrightarrow\quad \mathbf{d} = \mathbf{a}+2\mathbf{b}$$

and **a**, **b**, and **d** are said to be *linearly dependent* on each other (as opposed to being *linearly independent* of each other).

To find a normal to the plane, **n**, we take the cross product of any two of the coplanar vectors, say $\mathbf{a}\times\mathbf{b}$. With $\mathbf{n} = (2,5,-4)$, the equation of the plane is

$$\mathbf{r}\bullet(2,5,-4) = \beta$$

where $\mathbf{r} = (x,y,z)$ is a point in the plane and $\beta$ is a (scalar) constant. Putting $\mathbf{r}=\mathbf{a}$, for example, we find that $\beta = (1,2,3)\bullet(2,5,-4) = 0$. Hence

$$\underline{\mathbf{r}\bullet(2,5,-4) = 0} \quad \text{or} \quad \underline{2x+5y-4z = 0}$$

The perpendicular distance of the origin to the plane is $\underline{zero}$ because it is given by the right-hand side of the vector equation for a plane when the normal vector has unit length. This can be achieved by dividing both sides of the equation by $|(2,5,-4)|$ which, of course, still leaves zero on the right-hand side.

---

**(9)** Given three non-coplanar vectors **a**, **b**, and **c**, a corresponding set of 'reciprocal vectors' are defined by

$$A = (b \times c)/s, \quad B = (c \times a)/s \quad \text{and} \quad C = (a \times b)/s$$

where $s = \mathbf{a} \cdot (\mathbf{b} \times \mathbf{c})$. Show that $\mathbf{a} \cdot \mathbf{A} = 1$, $\mathbf{a} \cdot \mathbf{B} = \mathbf{a} \cdot \mathbf{C} = 0$ and find similar expressions for **b** and **c**. Hence, derive simple formulae for the coefficients $\alpha$, $\beta$, and $\gamma$ of a vector $\mathbf{r} = \alpha \mathbf{a} + \beta \mathbf{b} + \gamma \mathbf{c}$. Evaluate $\mathbf{A} \cdot (\mathbf{B} \times \mathbf{C})$ in terms of $s$.

---

Since the scalar triple product of coplanar vectors is zero,

$$\mathbf{a} \cdot \mathbf{A} = \mathbf{a} \cdot (\mathbf{b} \times \mathbf{c})/s = s/s = 1$$

$$\mathbf{a} \cdot \mathbf{B} = \mathbf{a} \cdot (\mathbf{c} \times \mathbf{a})/s = 0/s = 0$$

$$\mathbf{a} \cdot \mathbf{C} = \mathbf{a} \cdot (\mathbf{a} \times \mathbf{b})/s = 0/s = 0$$

$$[\mathbf{a}, \mathbf{b}, \mathbf{c}] = \mathbf{a} \cdot (\mathbf{b} \times \mathbf{c})$$
$$= \mathbf{b} \cdot (\mathbf{c} \times \mathbf{a})$$
$$= \mathbf{c} \cdot (\mathbf{a} \times \mathbf{b})$$

As a scalar triple product is also invariant with respect to a cyclic permutation of the three vectors, it follows that

$$\mathbf{b} \cdot \mathbf{B} = \mathbf{b} \cdot (\mathbf{c} \times \mathbf{a})/s = 1 \quad \text{and} \quad \mathbf{c} \cdot \mathbf{C} = \mathbf{c} \cdot (\mathbf{a} \times \mathbf{b})/s = 1$$

whereas

$$\mathbf{b} \cdot \mathbf{A} = \mathbf{b} \cdot \mathbf{C} = \mathbf{c} \cdot \mathbf{A} = \mathbf{c} \cdot \mathbf{B} = 0/s = 0$$

If a vector **r** is expressed using **a**, **b**, and **c** as the basis vectors, with coefficients $\alpha$, $\beta$, and $\gamma$ respectively, then

$$\mathbf{r} \cdot \mathbf{A} = \alpha \mathbf{a} \cdot \mathbf{A} + \beta \mathbf{b} \cdot \mathbf{A} + \gamma \mathbf{c} \cdot \mathbf{A} = \alpha + 0 + 0$$

Therefore $\alpha = \mathbf{r} \cdot \mathbf{A}$ and, similarly, $\beta = \mathbf{r} \cdot \mathbf{B}$ and $\gamma = \mathbf{r} \cdot \mathbf{C}$.

The scalar triple product of the reciprocal vectors can be evaluated from their definitions by using eqn (8.16) to expand a vector triple product (when $\mathbf{c} \times \mathbf{a}$, or $\mathbf{a} \times \mathbf{b}$, is treated as a single entity):

$$\mathbf{A} \cdot (\mathbf{B} \times \mathbf{C}) = (\mathbf{b} \times \mathbf{c}) \cdot \{(\mathbf{c} \times \mathbf{a}) \times (\mathbf{a} \times \mathbf{b})\}/s^3$$

$$= (\mathbf{b} \times \mathbf{c}) \cdot \{\underbrace{[(\mathbf{c} \times \mathbf{a}) \cdot \mathbf{b}]}_{s}\mathbf{a} - \underbrace{[(\mathbf{c} \times \mathbf{a}) \cdot \mathbf{a}]}_{0}\mathbf{b}\}/s^3$$

$$= (\mathbf{b} \times \mathbf{c}) \cdot \mathbf{a}/s^2$$

$$= s/s^2 = 1/s$$

Reciprocal vectors are often used to simplify problems in crystallography and solid-state physics and chemistry.

# 9 Matrices

## 9.1 Definition and nomenclature

In this chapter we will be considering the topic of *matrices*. Although our discussion may seem rather abstract at the outset, it is an important topic that often goes under the heading of *linear algebra*.

The simplest definition of a matrix is that it is a rectangular array of numbers, usually enclosed in large round brackets. If it consists of M rows and N columns, it is said to be an $M \times N$ matrix. For example

$$\mathbb{A} = \begin{pmatrix} 2 & 5 & 3 \\ 1 & 4 & 7 \end{pmatrix} \tag{9.1}$$

is a $2 \times 3$ matrix. We will use double-stroked letters for matrices, to distinguish them from scalars and vectors. If a matrix is denoted by $\mathbb{A}$, as in eqn (9.1), then its individual *elements* are specified by appending two suffices, $i$ and $j$, to A that give their locations in terms of row and column respectively: $A_{ij}$, where $i = 1$ or 2 and $j = 1$, 2, or 3 in the case of eqn (9.1). Thus $A_{11} = 2$, $A_{21} = 1$, $A_{12} = 5$, and so on.

$$\mathbb{A} = \begin{pmatrix} A_{11} & A_{12} & A_{13} & \cdots & A_{1N} \\ A_{21} & A_{22} & A_{23} & \cdots & A_{2N} \\ \vdots & \vdots & \vdots & & \vdots \\ A_{M1} & A_{M2} & A_{M3} & \cdots & A_{MN} \end{pmatrix}$$

Depending on their size (or shape), and the properties of their elements, matrices are often given qualifying names. For example, a *row* matrix has only one row (M = 1) and a *column* matrix consists of a single column (N = 1); in fact these can be considered to be vectors, requiring just one suffix, with the elements being the relevant components. If the number of rows and columns are equal (M = N), then it is said to be a *square* matrix; this case is of special interest, and the bulk of this chapter is devoted to its study.

Two properties of matrices which are frequently met in practice is that they are *real* and *symmetric*. The former means the elements are not complex numbers and the second, which implicitly assumes a square matrix, that the matrix remains the same if its rows and columns are interchanged. The operation of switching around rows and columns is called a *transpose*, and is denoted by attaching a 'T' superscript to the matrix; thus we can formally write that the $ij^{\text{th}}$ element of $\mathbb{A}^{\mathsf{T}}$ is equal to the $ji^{\text{th}}$ element of $\mathbb{A}$:

$$\mathbb{A}^{\mathsf{T}} = \begin{pmatrix} A_{11} & A_{21} & \cdots & A_{M1} \\ A_{12} & A_{22} & \cdots & A_{M2} \\ A_{13} & A_{23} & \cdots & A_{M3} \\ \vdots & \vdots & & \vdots \\ A_{1N} & A_{2N} & \cdots & A_{MN} \end{pmatrix}$$

$$\left( \mathbb{A}^{\mathsf{T}} \right)_{ij} = \left( \mathbb{A} \right)_{ji} = A_{ji} \tag{9.2}$$

$$\mathbb{A}^{\mathsf{T}} \equiv \widetilde{\mathbb{A}}$$

The transpose of a column matrix is, of course, a row matrix, and vice versa. A real and symmetric matrix is, therefore, defined by the properties that

$$\mathbb{A}^* = \mathbb{A} \quad \text{and} \quad \mathbb{A}^\mathsf{T} = \mathbb{A} \tag{9.3}$$

where the superscripted '*' corresponds to the operation of replacing each element with its complex conjugate. A type of matrix which plays a central rôle in quantum mechanics is called *Hermitian*; it satisfies the condition that

$$\mathbb{A}^\mathsf{T} = \mathbb{A}^* \tag{9.4}$$

A consideration of eqns (9.3) and (9.4) shows that a real and symmetric matrix is a special case of a Hermitian one.

## 9.2 Matrix arithmetic

The sum or difference of two matrices, say $\mathbb{A}$ and $\mathbb{B}$, is calculated by adding or subtracting the corresponding elements

$$\mathbb{C} = \mathbb{A} \pm \mathbb{B} \quad \Longleftrightarrow \quad c_{ij} = A_{ij} \pm B_{ij} \tag{9.5}$$

$$\begin{pmatrix} 2 & 5 \\ 1 & 4 \\ 6 & 1 \end{pmatrix} + \begin{pmatrix} 3 & -1 \\ 7 & 0 \\ -3 & 1 \end{pmatrix} = \begin{pmatrix} 5 & 4 \\ 8 & 4 \\ 3 & 2 \end{pmatrix}$$

Both matrices must be of the same size for this combination to be viable; in other words, if $\mathbb{A}$ is an $N \times M$ matrix so too must be $\mathbb{B}$ (and $\mathbb{C}$).

Matrix multiplication is far less straightforward than addition and subtraction. It is defined as follows: the $ij^{\text{th}}$ element of $\mathbb{A}\mathbb{B}$ is given by the dot product of the $i^{\text{th}}$ row of $\mathbb{A}$ with the $j^{\text{th}}$ column of $\mathbb{B}$. Explicitly,

$$\mathbb{C} = \mathbb{A}\mathbb{B} \quad \Longleftrightarrow \quad c_{ij} = \sum_k A_{ik} B_{kj} \tag{9.6}$$

$$\begin{pmatrix} 2 & 5 \\ 1 & 4 \\ 6 & 1 \end{pmatrix} \begin{pmatrix} 3 & -1 \\ 7 & 0 \end{pmatrix} = \begin{pmatrix} 41 & -2 \\ 31 & -1 \\ 25 & -6 \end{pmatrix}$$

where the summation over the index $k$ requires that $\mathbb{A}$ has the same number of columns as $\mathbb{B}$ has rows. The usefulness of the rule of eqn (9.6) will become clearer later, but we should note that it leads to an algebra where multiplication is no longer commutative; that is to say

$$\mathbb{A}\mathbb{B} \neq \mathbb{B}\mathbb{A}$$

in general. Indeed, $\mathbb{B}\mathbb{A}$ may not even exist, even if $\mathbb{A}\mathbb{B}$ does, because of the size restrictions on the numbers of rows and columns needed for a product. This lack of commutativity means that we must be very careful in stating whether both sides of an equation are to be pre- or post-multiplied by a matrix – the order matters!

Incidentally, the transpose of a product is equal to the product of the transposes but in reverse order

$$\left(\mathbb{A}\mathbb{B}\right)^\mathsf{T} = \mathbb{B}^\mathsf{T}\mathbb{A}^\mathsf{T} \tag{9.7}$$

a result which can be derived from eqns (9.2) and (9.6).

Division is not allowed in matrix algebra, but we will shortly see how this limitation can sometimes be circumvented. The only exception is division by a scalar (or a $1\times1$ matrix) which, like its multiplication counterpart, is an anomalous case. It is a very simple operation

$$\mathbb{C} = \alpha\,\mathbb{B} \quad\Longleftrightarrow\quad C_{ij} = \alpha\,B_{ij} \tag{9.8}$$

where every element of the matrix is multiplied or divided by the constant $\alpha$.

$$2\begin{pmatrix} 3 & -1 \\ 7 & 0 \\ -3 & 1 \end{pmatrix} = \begin{pmatrix} 6 & -2 \\ 14 & 0 \\ -6 & 2 \end{pmatrix}$$

## 9.3 Determinants

Before continuing with our general study of matrices, we must make a brief digression to consider the topic of *determinants*. This is an essential prerequisite for much of the discussion about square matrices that is to follow in the rest of this chapter.

The easiest case is that of a $1\times1$ matrix, whose determinant is simply equal to the element itself: $\det(\mathbb{A}) = |A_{11}| = A_{11}$. Moving onto a less trivial situation, the determinant of a $2\times2$ matrix is defined to be

$$\det(\mathbb{A}) = \begin{vmatrix} A_{11} & A_{12} \\ A_{21} & A_{22} \end{vmatrix} = A_{11}A_{22} - A_{12}A_{21} \tag{9.9}$$

The explicit formulae for the determinants of higher-order matrices become increasingly complicated, but can always be related to ones of a lower-order with a straightforward rule: the determinant of a matrix is given by the dot product of any of its rows or columns with its corresponding *cofactors*. The cofactor of the matrix-element $A_{ij}$ is, in turn, given by $(-1)^{i+j}$ times the determinant of the matrix which remains when the $i^{\text{th}}$ row and $j^{\text{th}}$ column of $\mathbb{A}$ are removed; the former yields a checker-board scheme of $+1$ and $-1$, and the latter, known as the *minor*, is the determinant of a smaller matrix (by one) than the original.

$$\begin{vmatrix} 2 & 5 \\ 1 & 4 \end{vmatrix} = 2\times4 - 5\times1 = 3$$

$$\begin{pmatrix} + & - & + & \cdots \\ - & + & - & \cdots \\ + & - & + & \cdots \\ \vdots & \vdots & \vdots & \ddots \end{pmatrix}$$

Thus, for example, the determinant of a $3\times3$ matrix can be written in terms of the determinants of three $2\times2$ matrices

$$\begin{vmatrix} A_{11} & A_{12} & A_{13} \\ A_{21} & A_{22} & A_{23} \\ A_{31} & A_{32} & A_{33} \end{vmatrix} = A_{11}\begin{vmatrix} A_{22} & A_{23} \\ A_{32} & A_{33} \end{vmatrix} - A_{12}\begin{vmatrix} A_{21} & A_{23} \\ A_{31} & A_{33} \end{vmatrix} + A_{13}\begin{vmatrix} A_{21} & A_{22} \\ A_{31} & A_{32} \end{vmatrix}$$

where we have carried out the calculation with respect to the top row. Although unnecessary given eqn (9.9), the determinant of a $2\times2$ matrix can itself be written in terms of the determinants of two $1\times1$ matrices; this is a useful exercise to carry out (once), since it enables us to verify that the determinant yielded by the cofactor rule is independent of the choice of row or column.

$$\begin{vmatrix} 1 & 0 & 2 \\ 3 & -1 & 5 \\ 4 & 3 & 7 \end{vmatrix} = \begin{vmatrix} -1 & 5 \\ 3 & 7 \end{vmatrix} + 2\begin{vmatrix} 3 & -1 \\ 4 & 3 \end{vmatrix}$$

$$= -7 - 15 + 2\,(9+4)$$

$$= 4$$

Determinants have a certain number of general properties, some of which are useful in facilitating their evaluation. These are listed without proof, but can be made plausible by checking that they work explicitly for a $2\times2$ matrix: (i) interchanging rows with columns leaves the determinant unchanged; (ii) multiplying a row or column by a constant $\beta$ multiplies the determinant by $\beta$; (iii) if one row (or column) is zero then, then so too is the determinant; (iv) if one row (or column) is a multiple of another, then the determinant is zero; (v) interchanging

$$\det(\mathbb{A}^{\mathsf{T}}) = \det(\mathbb{A})$$

$$\det(\mathbb{AB}) = \det(\mathbb{A})\det(\mathbb{B})$$

two adjacent rows (or columns) multiplies the determinant by $-1$; (vi) the addition of a multiple of one row (or column) to another leaves the determinant unchanged; (vii) the determinant of a product of matrices is equal to the product of their determinants.

Incidentally, the cofactor rule for determinants provides a good method for remembering the formula for a cross product given in eqn (8.10)

$$\mathbf{a} \times \mathbf{b} = \begin{vmatrix} \mathbf{i} & \mathbf{j} & \mathbf{k} \\ a_x & a_y & a_z \\ b_x & b_y & b_z \end{vmatrix} \tag{9.10}$$

$$(\mathbf{a} \times \mathbf{b}) \cdot \mathbf{c} = \begin{vmatrix} c_x & c_y & c_z \\ a_x & a_y & a_z \\ b_x & b_y & b_z \end{vmatrix}$$

What's more, it's not too difficult to see that a scalar triple product is obtained if the top row is replaced by the components of a third vector $\mathbf{c}$. Thus, with reference to section 8.5, the determinant of a $3 \times 3$ matrix represents the volume of a parallelepiped. In fact, this is the general physical interpretation of a determinant — its magnitude gives the 'volume' of the object generated when the rows, or columns, are used as bounding vectors. This reduces to a simple length for the case of a $1 \times 1$ matrix, and the area of a parallelogram for a $2 \times 2$ one.

## 9.4 Inverse matrices

$$\mathbb{I} = \begin{pmatrix} 1 & 0 & 0 & \dots & 0 \\ 0 & 1 & 0 & \dots & 0 \\ 0 & 0 & 1 & & 0 \\ \vdots & \vdots & & \ddots & \vdots \\ 0 & 0 & 0 & \dots & 1 \end{pmatrix}$$

We said in section 9.2 that division by a matrix was not allowed. There is a manipulation for square matrices, however, that mimics this operation: multiplication by an *inverse* matrix. The latter is defined by the property that

$$\mathbb{A}^{-1}\mathbb{A} = \mathbb{A}\,\mathbb{A}^{-1} = \mathbb{I} \tag{9.11}$$

where $\mathbb{I}$ is the matrix equivalent of unity, consisting of ones down the diagonal and zeros everywhere else, and is called an *identity* or *unit* matrix; anything (permissible) multiplied by $\mathbb{I}$ returns the entity itself.

Although the inverse of a matrix could be ascertained from the definition of eqn (9.11), a more systematic approach is provided by the following result

$$\begin{pmatrix} 5 & 3 \\ 2 & 1 \end{pmatrix}^{-1} = \frac{1}{5-6}\begin{pmatrix} 1 & -3 \\ -2 & 5 \end{pmatrix}$$

$$= \begin{pmatrix} -1 & 3 \\ 2 & -5 \end{pmatrix}$$

$$\mathbb{A}^{-1} = \frac{\text{adj}(\mathbb{A})}{\det(\mathbb{A})} \tag{9.12}$$

where $\text{adj}(\mathbb{A})$ is the *adjoint* of $\mathbb{A}$; it is a matrix where each of the elements of the transpose of $\mathbb{A}$, $\mathbb{A}^{\mathsf{T}}$, has been replaced by its cofactor. The denominator in eqn (9.12) indicates that the inverse of $\mathbb{A}$, $\mathbb{A}^{-1}$, does not exist if its determinant, $\det(\mathbb{A})$, is equal to zero; such a matrix is said to be *singular*.

## 9.5 Linear simultaneous equations

We met simultaneous equations in section 1.4, and noted that the easiest ones to solve are those which are linear. In the context of the current chapter, these are ones that can be cast in the following general terms

$$\mathbb{A}\mathbf{x} = \mathbf{b} \tag{9.13}$$

where $\mathbb{A}$ is a matrix, and **x** and **b** are vectors or column matrices; the elements of $\mathbb{A}$ and **b** are known whereas those of **x**, which satisfy eqn (9.13), are sought. For example, the pair of equations in section 1.4 can be written as

$$\begin{pmatrix} a & b \\ c & d \end{pmatrix}\begin{pmatrix} x \\ y \end{pmatrix} = \begin{pmatrix} \alpha \\ \beta \end{pmatrix}$$

where $a$, $b$, $c$, $d$, $\alpha$, and $\beta$ are constants.

A formula for the solution of eqn (9.13) is readily obtained if both sides are premultiplied by the inverse of $\mathbb{A}$:

$$\mathbb{A}^{-1}\mathbb{A}\,\mathbf{x} = \mathbb{A}^{-1}\mathbf{b} \quad \Rightarrow \quad \mathbb{I}\,\mathbf{x} = \mathbf{x} = \mathbb{A}^{-1}\mathbf{b} \qquad (9.14)$$

where we have used the property of eqn (9.11), and that of the identity matrix. The evaluation of the inverse may require some effort, of course, but the actual prescription of eqn (9.14) for finding **x** is straightforward.

From our discussion in section 9.4, it is clear that we have implicitly assumed that: (i) the matrix $\mathbb{A}$ is square; (ii) $\det(\mathbb{A}) \neq 0$. The first point means that there must be as many simultaneous equations as unknowns, and the second ensures that they are *linearly independent*. That is to say we can't construct any one of them from a simple combination of the others, so that there are really fewer equations than it would appear. If $\mathbb{A}$ is singular, then either the solution is not unique or there is no solution at all. Geometrically, for the case of a 2×2 matrix, we seek the intersection of two straight lines. They will not meet at a single point if they are parallel; if they lie on top of each other, then everywhere along the line is a solution; otherwise, the simultaneous conditions are never satisfied.

$$\begin{pmatrix} x \\ y \end{pmatrix} = \begin{pmatrix} a & b \\ c & d \end{pmatrix}^{-1}\begin{pmatrix} \alpha \\ \beta \end{pmatrix}$$

$$= \tfrac{1}{\Delta}\begin{pmatrix} d & -b \\ -c & a \end{pmatrix}\begin{pmatrix} \alpha \\ \beta \end{pmatrix}$$

$$\left[\text{For } \Delta = ad - bc \neq 0\right]$$

## 9.6 Transformations

When a matrix (not necessarily square) is applied to a vector, it generates a new column matrix (or vector)

$$\mathbb{A}\mathbf{x} = \mathbf{y} \qquad (9.15)$$

The change from the 'input' **x** to the 'output' **y** is called a *transformation*, or a *mapping*, with the matrix $\mathbb{A}$ often referred to as an *operator*. For example, when a two-dimensional vector is premultiplied by

$$\mathbb{A} = \begin{pmatrix} \cos\theta & -\sin\theta \\ \sin\theta & \cos\theta \end{pmatrix} \qquad (9.16)$$

it is rotated anticlockwise through an angle $\theta$. As with all 2 × 2 matrices, the nature of the transformation is best ascertained by considering its effect on the corners of the unit square. That is, by working out where the following points map to: $(0,0)$, $(0,1)$, $(1,0)$, and $(1,1)$. If $\mathbb{A}$ is square, and not singular, then a reverse transformation is obtained by applying the inverse, $\mathbb{A}^{-1}$, to **y**.

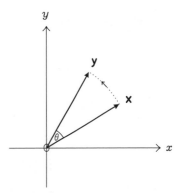

## 9.7 Eigenvalues and eigenvectors

There is a special case of eqn (9.15), $\mathbf{y} = \lambda\mathbf{x}$, called the *eigenvalue* equation, which is of great physical importance. This is when $\mathbf{x}$ maps onto itself, give or take a scale factor $\lambda$:

$$\mathbb{A}\,\mathbf{x} = \lambda\mathbf{x} \qquad (9.17)$$

The values of $\lambda$ and $\mathbf{x}$ which satisfy this equation are referred to as the *eigenvalues* and *eigenvectors* of $\mathbb{A}$ respectively.

The matrix $\mathbb{A}$ here is implicitly square, so eqn (9.17) becomes

$$\left(\mathbb{A} - \lambda\mathbb{I}\right)\mathbf{x} = \mathbf{0} \qquad (9.18)$$

where we have made use of the property that $\mathbb{I}\mathbf{x} = \mathbf{x}$, and the unit matrix is of the same size as $\mathbb{A}$. Following our discussion in section 9.5, premultiplication of both sides of eqn (9.18) by the inverse of the composite matrix $(\mathbb{A} - \lambda\mathbb{I})$ leads to the conclusion that $\mathbf{x} = \mathbf{0}$. The only way we can escape this fate is if

$$\det\left(\mathbb{A} - \lambda\mathbb{I}\right) = 0 \qquad (9.19)$$

for then $(\mathbb{A} - \lambda\mathbb{I})^{-1}$ does not exist, and so we are not forced to accept the *trivial solution* of $\mathbf{x} = \mathbf{0}$.

$$\begin{vmatrix} 4-\lambda & 1 \\ 2 & 3-\lambda \end{vmatrix} = 0$$

$$\Rightarrow \lambda^2 - 7\lambda + 10 = 0$$

$$\therefore \ \lambda = 2 \ \text{ or } \ \lambda = 5$$

The use of eqn (9.19) is usually the first step in solving an eigenvalue equation. It tells us to subtract $\lambda$ from each of the diagonal elements of $\mathbb{A}$, and to set the determinant of the resulting matrix equal to zero. For an $N \times N$ matrix, this gives rise to the problem of finding the roots of an $N^{\text{th}}$-order polynomial in $\lambda$ (called the *characteristic* equation). Thus there are N eigenvalues $(\lambda_1, \lambda_2, \lambda_3, \ldots, \lambda_N)$, and we have to consider each in turn to find the corresponding eigenvectors $(\mathbf{x}_1, \mathbf{x}_2, \mathbf{x}_3, \ldots, \mathbf{x}_N)$.

When $\lambda = 2$, $\ 2x + y = 0$

$$\therefore \ \mathbf{x} = \alpha \begin{pmatrix} 1 \\ -2 \end{pmatrix}$$

To work out the eigenvectors we substitute for the $\lambda$, one at a time, into eqn (9.17) or eqn (9.18). This does not yield a unique solution for the corresponding $\mathbf{x}$, because of eqn (9.19), but it does specify certain relationships between its components that must be satisfied. In the simplest case, we are left with just one unknown in the problem, say $\beta$, that determines the length of $\mathbf{x}$ but its direction is fixed; $\beta$ is often chosen so that the eigenvector is *normalized* ($|\mathbf{x}| = 1$). For the

When $\lambda = 5$, $\ -x + y = 0$

$$\therefore \ \mathbf{x} = \beta \begin{pmatrix} 1 \\ 1 \end{pmatrix}$$

*degenerate* case, when there are repeated roots of the characteristic equation for the eigenvalues, the directions of the corresponding eigenvectors are not fixed. If a value of $\lambda$ occurs twice, for example, then any $\mathbf{x}$ that lies in a specific plane will be an eigenvector; we are free to choose any two linearly independent directions in the plane as the relevant eigenvectors.

In problems of physical interest, the matrices that are encountered tend to be real and symmetric or, more generally, Hermitian; these are defined by eqns (9.3) and (9.4) respectively. It can be shown that such matrices have very convenient and pleasant eigen-properties. Namely, that all the eigenvalues are real and the eigenvectors are mutually orthogonal:

$$\mathbb{A}\,\mathbf{x}_j = \lambda_j\,\mathbf{x}_j \qquad\qquad \lambda_j = \lambda_j^* \quad \text{and} \quad \mathbf{x}_j^{\mathsf{T}}\mathbf{x}_k = \mathbf{x}_j \bullet \mathbf{x}_k = 0 \ \text{ if } \ j \neq k \qquad (9.20)$$

where the subscripts label the different solutions of eqn (9.17). What's more, the product of the eigenvalues turns out to be equal to $\det(\mathbb{A})$ and their sum is the sum of the diagonal elements, or *trace*, of $\mathbb{A}$ ($\Sigma A_{ii}$).

$$\det(\mathbb{A}) = \lambda_1 \lambda_2 \lambda_3 \cdots \lambda_N$$

$$\text{trace}(\mathbb{A}) = \lambda_1 + \lambda_2 + \cdots + \lambda_N$$

## 9.8 Diagonalization

We can obtain a geometrical interpretation of eigenvalues and eigenvectors by considering the following scalar quantity

$$Q = \mathbf{x}^T \mathbb{A} \mathbf{x} \tag{9.21}$$

which is called a *quadratic form*. If $\mathbb{A}$ is a real and symmetric $2 \times 2$ matrix, for example, with both eigenvalues having the same sign, and the vector $\mathbf{x}^T$ is the row matrix $(x \ \ y)$, then eqn (9.21) yields the formula for an ellipse. The resulting equation won't be quite as simple as eqn (2.10), in general, because the ellipse will be skew with respect to the $x$ and $y$ axes. In this picture, the eigenvectors are the directions of the principal axes and the eigenvalues are inversely proportional to the squares of the corresponding widths. Higher-order matrices give rise to multidimensional ellipsoids, but the principal axes relationship to the eigen-properties still holds.

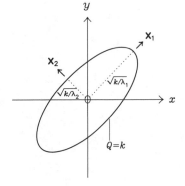

Having evaluated the eigenvalues and eigenvectors of $\mathbb{A}$, any subsequent analysis can often be simplified by working in a new set of coordinates, $\mathbf{u}$, that are aligned with the principal axes of the ellipse. The required transformation between $\mathbf{x}$ and $\mathbf{u}$ is given by

$$\mathbf{x} = \mathbb{O}\mathbf{u} \tag{9.22}$$

where the columns of the matrix $\mathbb{O}$ consist of the normalized eigenvectors of $\mathbb{A}$; in conjunction with eqn (9.20), therefore, we find that the mapping matrix is *orthogonal* in that it satisfies the condition

$$\mathbb{O}^T \mathbb{O} = \mathbb{I} \tag{9.23}$$

$$\mathbb{O} = \begin{pmatrix} | & | & & | \\ \hat{\mathbf{x}}_1 & \hat{\mathbf{x}}_2 & \cdots & \hat{\mathbf{x}}_N \\ | & | & & | \end{pmatrix}$$

With eqn (9.22), the quadratic form of eqn (9.21) reduces to

$$Q = \mathbf{u}^T \mathbb{D} \mathbf{u} \tag{9.24}$$

$$\mathbb{D} = \mathbb{O}^T \mathbb{A} \mathbb{O}$$

where $\mathbb{D}$ is a *diagonal* matrix: it has zeros everywhere except down the diagonal, where the elements are equal to the eigenvalues of $\mathbb{A}$.

The physical value of an eigen-analysis is that it decomposes apparently complicated behaviour into its basic constituent parts. The eigenvectors associated with molecular vibrations, for example, correspond to the *normal modes* of the system, and the eigenvalues give their natural frequencies. Similarly, the eigenvectors and eigenvalues encountered in a quantum mechanics problem relate to the wave functions of the *stationary states* and their energy levels.

Incidentally, if two eigenvalues are equal ($\lambda_1 = \lambda_2$, say) then the quadratic form is a circle. Since this does not have distinct principal axes, it confirms that every direction in the relevant plane is an eigenvector in the degenerate case. As any

$$= \begin{pmatrix} \lambda_1 & 0 & 0 & \cdots & 0 \\ 0 & \lambda_2 & 0 & \cdots & 0 \\ 0 & 0 & \lambda_3 & \cdots & 0 \\ \vdots & \vdots & \vdots & \ddots & \vdots \\ 0 & 0 & 0 & \cdots & \lambda_N \end{pmatrix}$$

two-dimensional vector can be constructed from two basis vectors, we are at liberty to select any two independent directions as being the eigenvectors; by convention, they are chosen to be orthogonal.

## 9.9 Worked examples

**(1)** Find $\mathbb{A}+\mathbb{B}$, $\mathbb{A}-\mathbb{B}$, $\mathbb{A}\mathbb{B}$ and $\mathbb{B}\mathbb{A}$ when

$$\mathbb{A} = \begin{pmatrix} 2 & 1 \\ 1 & 2 \end{pmatrix} \quad \text{and} \quad \mathbb{B} = \begin{pmatrix} 3 & 3 \\ 0 & 4 \end{pmatrix}$$

Verify that $(\mathbb{A}\mathbb{B})^{\mathsf{T}} = \mathbb{B}^{\mathsf{T}}\mathbb{A}^{\mathsf{T}}$ and that $\det(\mathbb{A}\mathbb{B}) = \det(\mathbb{A})\det(\mathbb{B})$.

$$\mathbb{A}+\mathbb{B} = \begin{pmatrix} 2+3 & 1+3 \\ 1+0 & 2+4 \end{pmatrix} = \begin{pmatrix} 5 & 4 \\ 1 & 6 \end{pmatrix}$$

$$\mathbb{A}-\mathbb{B} = \begin{pmatrix} 2-3 & 1-3 \\ 1-0 & 2-4 \end{pmatrix} = \begin{pmatrix} -1 & -2 \\ 1 & -2 \end{pmatrix}$$

$$\mathbb{A}\mathbb{B} = \begin{pmatrix} 2\times3+1\times0 & 2\times3+1\times4 \\ 1\times3+2\times0 & 1\times3+2\times4 \end{pmatrix} = \begin{pmatrix} 6 & 10 \\ 3 & 11 \end{pmatrix}$$

$$\mathbb{B}\mathbb{A} = \begin{pmatrix} 3\times2+3\times1 & 3\times1+3\times2 \\ 0\times2+4\times1 & 0\times1+4\times2 \end{pmatrix} = \begin{pmatrix} 9 & 9 \\ 4 & 8 \end{pmatrix}$$

This simple example illustrates that matrix multiplication is not commutative: $\mathbb{A}\mathbb{B} \neq \mathbb{B}\mathbb{A}$ in general, even when both products exist (or are permissible).

$$\mathbb{B}^{\mathsf{T}}\mathbb{A}^{\mathsf{T}} = \begin{pmatrix} 3 & 0 \\ 3 & 4 \end{pmatrix}\begin{pmatrix} 2 & 1 \\ 1 & 2 \end{pmatrix} = \begin{pmatrix} 6 & 3 \\ 10 & 11 \end{pmatrix} = \begin{pmatrix} 6 & 10 \\ 3 & 11 \end{pmatrix}^{\mathsf{T}} = (\mathbb{A}\mathbb{B})^{\mathsf{T}}$$

$$\det(\mathbb{A}) = \begin{vmatrix} 2 & 1 \\ 1 & 2 \end{vmatrix} = 2\times2 - 1\times1 = 3$$

$$\det(\mathbb{B}) = \begin{vmatrix} 3 & 3 \\ 0 & 4 \end{vmatrix} = 3\times4 - 0\times3 = 12$$

$$\det(\mathbb{A}\mathbb{B}) = \begin{vmatrix} 6 & 10 \\ 3 & 11 \end{vmatrix} = 6\times11 - 3\times10 = 36 = \det(\mathbb{A})\det(\mathbb{B})$$

**(2)** Find the smallest seven integers, $x_1$ to $x_7$, that balance the following chemical equation:

$$x_1 \, MnS + x_2 \, As_2 Cr_{10} O_{35} + x_3 \, H_2 SO_4$$
$$\longrightarrow x_4 \, HMnO_4 + x_5 \, AsH_3 + x_6 \, CrS_3O_{12} + x_7 \, H_2O$$

Following question 10 in section 1.6 of
**Linear Algebra and its applications**
by *David C. Lay* (2011, 4$^{th}$ ed) Pearson

To balance the equation, we need to ensure that the number of atoms of each type is the same on both sides:

$$
\begin{array}{ll}
\text{H} & 2x_3 = x_4 + 3x_5 + 2x_7 \\
\text{O} & 35x_2 + 4x_3 = 4x_4 + 12x_6 + x_7 \\
\text{S} & x_1 + x_3 = 3x_6 \\
\text{Mn} & x_1 = x_4 \\
\text{Cr} & 10x_2 = x_6 \\
\text{As} & 2x_2 = x_5
\end{array}
$$

With only six linear equations for seven unknowns, the solution is not unique. This is to be expected, and convenient, because we need one degree of freedom to be able to scale $x_1$ to $x_7$ (multiplicatively) to obtain the smallest possible set of integers. Letting the scaling variable be $x_1 = \alpha$, the equations above can be written in matrix form as

$$
\begin{pmatrix}
0 & -2 & 1 & 3 & 0 & 2 \\
35 & 4 & -4 & 0 & -12 & -1 \\
0 & -1 & 0 & 0 & 3 & 0 \\
0 & 0 & 1 & 0 & 0 & 0 \\
10 & 0 & 0 & 0 & -1 & 0 \\
2 & 0 & 0 & -1 & 0 & 0
\end{pmatrix}
\begin{pmatrix}
x_2 \\ x_3 \\ x_4 \\ x_5 \\ x_6 \\ x_7
\end{pmatrix}
=
\begin{pmatrix}
0 \\ 0 \\ \alpha \\ \alpha \\ 0 \\ 0
\end{pmatrix}
$$

Dealing with this $6 \times 6$ matrix by hand, through inversion or otherwise, is rather tedious, and it is better to reduce the size of the problem by explicitly making use of the simpler relationships between the coefficients:

$$x_4 = \alpha, \quad x_5 = 2x_2, \quad x_6 = 10x_2 \quad \text{and} \quad x_3 = 30x_2 - \alpha$$

so that the remaining two equations (for hydrogen and oxygen) become

$$
\begin{aligned}
54x_2 - 2x_7 &= 3\alpha \\
35x_2 - x_7 &= 8\alpha
\end{aligned}
\quad \Longleftrightarrow \quad
\begin{pmatrix} 54 & -2 \\ 35 & -1 \end{pmatrix}
\begin{pmatrix} x_2 \\ x_7 \end{pmatrix}
=
\begin{pmatrix} 3\alpha \\ 8\alpha \end{pmatrix}
$$

These can be solved with the matrix manipulation of section 9.5, or the elementary approach of section 1.4, and lead to the result

$$x_2 = \tfrac{13}{16}\alpha \quad \text{and} \quad x_7 = \tfrac{327}{16}\alpha$$

and, hence,

$$x_1 = x_4 = \alpha, \quad x_5 = \tfrac{26}{16}\alpha, \quad x_6 = \tfrac{130}{16}\alpha \quad \text{and} \quad x_3 = \tfrac{374}{16}\alpha$$

$$
\begin{aligned}
\begin{pmatrix} x_2 \\ x_7 \end{pmatrix}
&=
\begin{pmatrix} 54 & -2 \\ 35 & -1 \end{pmatrix}^{-1}
\begin{pmatrix} 3\alpha \\ 8\alpha \end{pmatrix} \\
&= \tfrac{1}{16}
\begin{pmatrix} -1 & 2 \\ -35 & 54 \end{pmatrix}
\begin{pmatrix} 3\alpha \\ 8\alpha \end{pmatrix}
\end{aligned}
$$

The smallest set of integers is obtained with $\alpha = 16$:

$$\underline{x_1 = 16, \ x_2 = 13, \ x_3 = 374, \ x_4 = 16, \ x_5 = 26, \ x_6 = 130, \ x_7 = 327}$$

**(3)** Find the eigenvalues and the normalized eigenvectors of

$$\mathbb{A} = \begin{pmatrix} 1 & 0 & 1 \\ 0 & -1 & 0 \\ 1 & 0 & 1 \end{pmatrix}$$

Confirm that the eigenvectors are mutually orthogonal, the sum of the eigenvalues is equal to the trace of $\mathbb{A}$ and the product of eigenvalues is equal to $\det(\mathbb{A})$. Construct the diagonalization matrix $\mathbb{O}$ and verify that $\mathbb{O}\,\mathbb{O}^\mathsf{T} = \mathbb{O}^\mathsf{T}\mathbb{O} = \mathbb{I}$. Check that the *similarity* transform $\mathbb{O}^\mathsf{T}\mathbb{A}\,\mathbb{O} = \mathbb{D}$ yields a diagonal matrix simply related to the eigenvalues of $\mathbb{A}$.

$$\mathbf{x} = \begin{pmatrix} x \\ y \\ z \end{pmatrix}$$

Eigenvalue equation: $\qquad \mathbb{A}\mathbf{x} = \lambda\mathbf{x} \quad \Rightarrow \quad (\mathbb{A} - \lambda\mathbb{I})\mathbf{x} = \mathbf{0}$

For non-trivial solutions, $\det(\mathbb{A} - \lambda\mathbb{I}) = 0$

$$\therefore \quad \begin{vmatrix} 1-\lambda & 0 & 1 \\ 0 & -1-\lambda & 0 \\ 1 & 0 & 1-\lambda \end{vmatrix} = (-1-\lambda)\begin{vmatrix} 1-\lambda & 1 \\ 1 & 1-\lambda \end{vmatrix} = 0$$

$$\therefore \quad (1+\lambda)\left(\lambda^2 - 2\lambda + 1 - 1\right) = 0$$

$$\therefore \quad \lambda(1+\lambda)(\lambda - 2) = 0$$

i.e.  Eigenvalues are  $\underline{\lambda = 0, \quad \lambda = -1, \quad \text{and} \quad \lambda = 2}$

$\lambda_1 = 0$ 
When $\lambda = 0$, $\qquad \left.\begin{array}{r} x + z = 0 \\ y = 0 \end{array}\right\} \quad \mathbf{x} = \begin{pmatrix} \alpha \\ 0 \\ -\alpha \end{pmatrix} \quad \Rightarrow \quad \underline{\widehat{\mathbf{x}}_1 = \tfrac{1}{\sqrt{2}}\begin{pmatrix} 1 \\ 0 \\ -1 \end{pmatrix}}$

$\lambda_2 = -1$ 
When $\lambda = -1$, $\qquad \left.\begin{array}{r} 2x + z = 0 \\ x + 2z = 0 \end{array}\right\} \quad \mathbf{x} = \begin{pmatrix} 0 \\ \beta \\ 0 \end{pmatrix} \quad \Rightarrow \quad \underline{\widehat{\mathbf{x}}_2 = \begin{pmatrix} 0 \\ 1 \\ 0 \end{pmatrix}}$

$\lambda_3 = 2$ 
When $\lambda = 2$, $\qquad \left.\begin{array}{r} -x + z = 0 \\ -3y = 0 \end{array}\right\} \quad \mathbf{x} = \begin{pmatrix} \gamma \\ 0 \\ \gamma \end{pmatrix} \quad \Rightarrow \quad \underline{\widehat{\mathbf{x}}_3 = \tfrac{1}{\sqrt{2}}\begin{pmatrix} 1 \\ 0 \\ 1 \end{pmatrix}}$

$$\mathbf{x}_1 \bullet \mathbf{x}_2 = \mathbf{x}_1^\mathsf{T}\mathbf{x}_2 = \alpha \times 0 + 0 \times \beta + (-\alpha) \times 0 = 0$$

Similarly, $\qquad \mathbf{x}_1^\mathsf{T}\mathbf{x}_3 = 0 \quad \text{and} \quad \mathbf{x}_2^\mathsf{T}\mathbf{x}_3 = 0$

$$\therefore \quad \underline{\text{Eigenvectors are mutually orthogonal}}$$

$$A_{11} + A_{22} + A_{33} = 1 \atop \lambda_1 + \lambda_2 + \lambda_3 = 1 \Bigg\} \quad \therefore \ \underline{\text{Sum of eigenvalues} = \text{trace}(A)} \qquad \text{trace}(A) = \sum_{i=1}^{N} A_{ii}$$

$$\begin{vmatrix} 1 & 0 & 1 \\ 0 & -1 & 0 \\ 1 & 0 & 1 \end{vmatrix} = -1 \times (1-1) = 0 \atop \lambda_1 \lambda_2 \lambda_3 = 0 \times (-1) \times 2 = 0 \Bigg\} \quad \therefore \ \underline{\text{Product of eigenvalues} = \text{det}(A)}$$

$$\text{Diagonalization matrix}, \ \mathbb{O} = \begin{pmatrix} \frac{1}{\sqrt{2}} & 0 & \frac{1}{\sqrt{2}} \\ 0 & 1 & 0 \\ \frac{-1}{\sqrt{2}} & 0 & \frac{1}{\sqrt{2}} \end{pmatrix} = \frac{1}{\sqrt{2}} \begin{pmatrix} 1 & 0 & 1 \\ 0 & \sqrt{2} & 0 \\ -1 & 0 & 1 \end{pmatrix}$$

$$\therefore \ \mathbb{O}\mathbb{O}^{\mathsf{T}} = \frac{1}{2} \begin{pmatrix} 1 & 0 & 1 \\ 0 & \sqrt{2} & 0 \\ -1 & 0 & 1 \end{pmatrix} \begin{pmatrix} 1 & 0 & -1 \\ 0 & \sqrt{2} & 0 \\ 1 & 0 & 1 \end{pmatrix} = \frac{1}{2} \begin{pmatrix} 2 & 0 & 0 \\ 0 & 2 & 0 \\ 0 & 0 & 2 \end{pmatrix} = \mathbb{I}$$

$$\mathbb{O}^{\mathsf{T}}\mathbb{O} = \frac{1}{2} \begin{pmatrix} 1 & 0 & -1 \\ 0 & \sqrt{2} & 0 \\ 1 & 0 & 1 \end{pmatrix} \begin{pmatrix} 1 & 0 & 1 \\ 0 & \sqrt{2} & 0 \\ -1 & 0 & 1 \end{pmatrix} = \frac{1}{2} \begin{pmatrix} 2 & 0 & 0 \\ 0 & 2 & 0 \\ 0 & 0 & 2 \end{pmatrix} = \mathbb{I}$$

$$\therefore \ \mathbb{O} \text{ is orthogonal, because } \ \underline{\mathbb{O}\mathbb{O}^{\mathsf{T}} = \mathbb{O}^{\mathsf{T}}\mathbb{O} = \mathbb{I}} \qquad \mathbb{O}^{\mathsf{T}} = \mathbb{O}^{-1}$$

$$\mathbb{O}^{\mathsf{T}}A = \frac{1}{\sqrt{2}} \begin{pmatrix} 1 & 0 & -1 \\ 0 & \sqrt{2} & 0 \\ 1 & 0 & 1 \end{pmatrix} \begin{pmatrix} 1 & 0 & 1 \\ 0 & -1 & 0 \\ 1 & 0 & 1 \end{pmatrix} = \frac{1}{\sqrt{2}} \begin{pmatrix} 0 & 0 & 0 \\ 0 & -\sqrt{2} & 0 \\ 2 & 0 & 2 \end{pmatrix}$$

$$\therefore \ \mathbb{O}^{\mathsf{T}}A\mathbb{O} = \frac{1}{2} \begin{pmatrix} 0 & 0 & 0 \\ 0 & -\sqrt{2} & 0 \\ 2 & 0 & 2 \end{pmatrix} \begin{pmatrix} 1 & 0 & 1 \\ 0 & \sqrt{2} & 0 \\ -1 & 0 & 1 \end{pmatrix} = \frac{1}{2} \begin{pmatrix} 0 & 0 & 0 \\ 0 & -2 & 0 \\ 0 & 0 & 4 \end{pmatrix}$$

$$\text{i.e.} \ \ \underline{\mathbb{O}^{\mathsf{T}}A\mathbb{O} = \begin{pmatrix} \lambda_1 & 0 & 0 \\ 0 & \lambda_2 & 0 \\ 0 & 0 & \lambda_3 \end{pmatrix} = \begin{pmatrix} 0 & 0 & 0 \\ 0 & -1 & 0 \\ 0 & 0 & 2 \end{pmatrix}}$$

When evaluating eigenvalues and eigenvectors, it is always worth checking that: (i) the sum of the eigenvalues is equal to the trace of the matrix; and (ii) the product of the eigenvalues is equal to the determinant of the matrix (if this is not too difficult to calculate). A failure to meet either of these requirements is indicative of a mistake, and should be corrected before the evaluation of the eigenvectors. For symmetric matrices, or more generally Hermitian ones, the eigenvalues must also be real ($\lambda = \lambda^*$) and the eigenvectors corresponding to distinct eigenvalues ($\lambda_i \neq \lambda_j$) should be orthogonal.

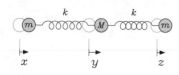

$$\frac{d^2x}{dt^2} = \ddot{x} = -w^2 x$$
$$\ddot{y} = -w^2 y$$
$$\ddot{z} = -w^2 z$$

**(4)** The longitudinal vibrational modes of carbon dioxide can be studied classically by modeling the linear molecule, $O = C = O$, as two equal masses $m$ attached with identical springs of stiffness $k$ to a central mass $M$. Denoting the displacements of $m$, $M$, and $m$ from their equilibrium positions along the bond direction by $x$, $y$, and $z$ respectively, and letting the frequency of the *normal modes* be $w$, Newton's Second Law of motion (and Hooke's law) tells us that the vibrating masses obey

$$\begin{pmatrix} mw^2 - k & k & 0 \\ k & Mw^2 - 2k & k \\ 0 & k & mw^2 - k \end{pmatrix} \begin{pmatrix} x \\ y \\ z \end{pmatrix} = \begin{pmatrix} 0 \\ 0 \\ 0 \end{pmatrix}$$

Find the frequencies of the normal modes (in terms of $m$, $M$, and $k$) and describe the motions of the atoms.

If this set of simultaneous equations is to have solutions other than the trivial $x = y = z = 0$, the $3 \times 3$ matrix must not have an inverse. This means that its determinant has to be zero:

$$\begin{vmatrix} mw^2 - k & k & 0 \\ k & Mw^2 - 2k & k \\ 0 & k & mw^2 - k \end{vmatrix} = 0$$

Evaluating the determinant,

$$\left( mw^2 - k \right) \left[ \left( mw^2 - k \right) \left( Mw^2 - 2k \right) - 2k^2 \right] = 0$$

which can be factorized as

$$w^2 \left( mw^2 - k \right) \left[ mMw^2 - k \left( 2m + M \right) \right] = 0$$

$$\ddot{x} = 0$$
$$\Rightarrow \quad x = vt + c$$

The $w = 0$ solution is not really of interest, because it corresponds to a steady drift of the molecule (*translational* mode): $x = y = z = vt + c$. The $w \neq 0$ solutions are the oscillation frequencies of the vibrational modes

$$w_1 = \sqrt{\frac{k}{m}} \quad \text{and} \quad w_2 = \sqrt{\frac{k \left( 2m + M \right)}{mM}}$$

Putting $w = w_1$ into the simultaneous equations gives

$$y = 0 \quad \text{and} \quad x = -z$$

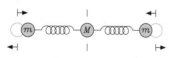

This is a symmetric stretching mode, where the carbon remains stationary and the two oxygen atoms move in opposite directions:

$$\ddot{x} = -w^2 x$$
$$\Rightarrow \quad x = A \sin(wt + \phi)$$

$$\begin{pmatrix} x \\ y \\ z \end{pmatrix} = \begin{pmatrix} 1 \\ 0 \\ -1 \end{pmatrix} A \sin(w_1 t + \phi)$$

Likewise, putting $\omega = \omega_2$ yields the asymmetric stretching mode

$$x = z \quad \text{and} \quad y = -\frac{2m}{M}x$$

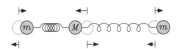

where the two oxygen atoms move in the same direction, by an equal amount, while the carbon goes the others way:

$$\begin{pmatrix} x \\ y \\ z \end{pmatrix} = \begin{pmatrix} 1 \\ -\frac{2m}{M} \\ 1 \end{pmatrix} B \sin(\omega_2 t + \psi)$$

Substituting $m=16$ and $M=12$, for oxygen and carbon, in the formulae for the normal frequencies yields $\omega_2/\omega_1 = \sqrt{11/3} \approx 1.9$. Measurements with *infrared* and *Raman* spectroscopy find a ratio of $\sim 1.8$, which is reasonable agreement given the simplicity of the model.

If this normal modes example seems similar to the analysis of an eigenvalue problem, it is because the equations of motion can easily be recast as an eigenvalue equation

$$\begin{pmatrix} -k/m & k/m & 0 \\ k/M & -2k/M & k/M \\ 0 & k/m & -k/m \end{pmatrix} \begin{pmatrix} x \\ y \\ z \end{pmatrix} = -\omega^2 \begin{pmatrix} x \\ y \\ z \end{pmatrix}$$

where $\lambda = -\omega^2$. It can be confirmed that the trace of this matrix is equal to the sum $0 - \omega_1^2 - \omega_2^2$, and that its determinant is zero.

---

**(5)** Show that the eigenvalues of a Hermitian matrix are real, and that the eigenvectors corresponding to distinct eigenvalues are orthogonal.

---

The equations for the $j^{\text{th}}$ and $k^{\text{th}}$ eigenvalues, $\lambda_j$ and $\lambda_k$, and the corresponding eigenvectors, $\mathbf{x}_j$ and $\mathbf{x}_k$, of a matrix $\mathbb{A}$ are

$$\mathbb{A}\,\mathbf{x}_j = \lambda_j\,\mathbf{x}_j \quad - (1) \qquad \text{and} \qquad \mathbb{A}\,\mathbf{x}_k = \lambda_k\,\mathbf{x}_k \quad - (2)$$

The transpose of eqn (1), and the complex conjugate of eqn (2), gives

$$(1)^{\mathsf{T}} \Rightarrow \qquad \mathbf{x}_j^{\mathsf{T}}\mathbb{A}^{\mathsf{T}} = \lambda_j\,\mathbf{x}_j^{\mathsf{T}} \quad - (3) \qquad\qquad (\mathbb{A}\,\mathbf{x})^{\mathsf{T}} = \mathbf{x}^{\mathsf{T}}\mathbb{A}^{\mathsf{T}}$$

$$(2)^{*} \Rightarrow \qquad \mathbb{A}^{*}\,\mathbf{x}_k^{*} = \lambda_k^{*}\,\mathbf{x}_k^{*} \quad - (4) \qquad\qquad \lambda^{\mathsf{T}} = \lambda$$

Subtracting eqn (4) pre-multiplied by $\mathbf{x}_j^{\mathsf{T}}$ from eqn (3) post-multiplied by $\mathbf{x}_k^{*}$, we obtain

$$(3)\mathbf{x}_k^{*} - \mathbf{x}_j^{\mathsf{T}}(4) \Rightarrow \qquad \mathbf{x}_j^{\mathsf{T}}(\mathbb{A}^{\mathsf{T}} - \mathbb{A}^{*})\,\mathbf{x}_k^{*} = (\lambda_j - \lambda_k^{*})\,\mathbf{x}_j^{\mathsf{T}}\mathbf{x}_k^{*}$$

$$\text{But} \quad \mathbb{A}^{\mathsf{T}} = \mathbb{A}^* \quad \Rightarrow \quad (\lambda_j - \lambda_k^*)\, \mathbf{x}_j^{\mathsf{T}} \mathbf{x}_k^* = 0 \quad \text{---} \ (5)$$

$$\mathbf{x}_j \neq 0 \qquad \text{If } j = k, \quad \underline{\lambda_j = \lambda_j^*} \quad \text{because} \quad \mathbf{x}_j^{\mathsf{T}} \mathbf{x}_j^* = |\mathbf{x}_j|^2 > 0$$

i.e.   The eigenvalues of a Hermitian matrix are real.

$\mathbf{a}^{\mathsf{T}}\mathbf{b}^*$ is a more general definition of a dot product than encountered earlier. It reduces to $\mathbf{a}^{\mathsf{T}}\mathbf{b}$, and hence eqn (8.7), when the components of vectors $\mathbf{a}$ and $\mathbf{b}$ are real, but ensures a real and positive result even if they are complex when $\mathbf{a} = \mathbf{b}$. This is in keeping with the interpretation of the scalar product of a vector with itself being the square its magnitude.

$$\text{If } j \neq k \text{ and } \lambda_j \neq \lambda_k, \quad (5) \Rightarrow \quad \underline{\mathbf{x}_j^{\mathsf{T}} \mathbf{x}_k^* = 0}$$

i.e.   The eigenvectors of distinct eigenvalues are orthogonal.

The condition $\mathbf{a}^{\mathsf{T}}\mathbf{b}^* = 0$ for $\mathbf{a} \neq \mathbf{b}$ (with $\mathbf{a} \neq 0$ and $\mathbf{b} \neq 0$) is the definition of orthogonality, so that the vectors $\mathbf{a}$ and $\mathbf{b}$ are perpendicular to each other.

If $\lambda_j = \lambda_k$, so that two (or more) eigenvalues are the same, then it is known as a *degenerate* case. The related eigenvectors do not then specify a unique direction, but correspond to a pair of vectors that lie in a given plane. Since any point in a plane can be attained through a linear combination of two (non-collinear) basis vectors lying in it, we are at liberty to choose two orthogonal directions in the degeneracy plane as being suitable eigenvectors. The preceding analysis applies equally well to symmetric matrices, $\mathbb{A}^{\mathsf{T}} = \mathbb{A}$, since these are just a special case of Hermitian ones (when $\mathbb{A}^* = \mathbb{A}$).

# Partial differentiation

**10**

## 10.1 Definition and the gradient vector

In chapter 4, on differentiation, we were concerned with learning about the rate at which two quantities varied with respect to each other. In practice, of course, most problems involve several entities; for example, a thermodynamic variable, such as the internal energy of a gas, will depend on temperature, pressure, and volume. The topic of *partial differentiation* is a natural extension of our earlier discussion.

To highlight the ideas clearly, and to avoid algebraic cluttering, we will focus principally on the case where a parameter $z$ is a function of just two variables, $x$ and $y$: $z = f(x, y)$. We can visualize such a situation from our everyday experience by thinking of $z$ as representing a height, with $x$ and $y$ being the two-dimensional coordinates of a point on the (flat) ground. While a realistic rendering of the topographical scene requires artistic ability, a more practical means of displaying the salient information is to draw a *contour map*; these are found in geographical atlases, where hills and valleys are depicted by a series of lines which join together regions of equal altitude.

Given the aforementioned setup, how can we convey a sense of the slope at any given point $(x, y)$? One possibility is to divide the problem into two parts where each is amenable to the analysis discussed in chapter 4: take slices of the $z$-surface along the $x$ and $y$ directions so that we obtain two sections, $z$ against $x$ (at fixed $y$) and $z$ versus $y$ (at constant $x$), which can be plotted as ordinary graphs. The pair of gradients which are generated in this way are called *partial derivatives*, because each provides some insight about the desired slope, and are denoted by curly d's

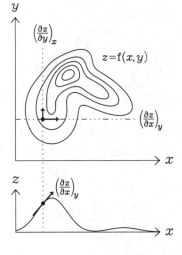

$$\left(\frac{\partial z}{\partial x}\right)_y = \lim_{\delta x \to 0} \frac{f(x + \delta x, y) - f(x, y)}{\delta x}$$

$$\left(\frac{\partial z}{\partial y}\right)_x = \lim_{\delta y \to 0} \frac{f(x, y + \delta y) - f(x, y)}{\delta y}$$

(10.1)

where the formulae constitute the appropriate generalization of eqn (4.1). Other notations for these partial derivatives include $\partial z/\partial x$ or $f_x(x, y)$, and $\partial z/\partial y$ or $f_y(x, y)$, but we recommend the form in eqn (10.1) where the quantity being held constant is specified explicitly (through a subscript).

Although eqn (10.1) yield a 'first principles' derivation of the partial derivatives of $z = f(x, y)$, they are normally evaluated by appealing to the results in chapter 4 while treating the relevant parameter as a constant. For example, the derivative of $z = 3x^2 + y^3$ with respect to $x$ at fixed $y$ is easily seen to be $(\partial z/\partial x)_y = 6x$; similarly, it's fairly obvious that $(\partial z/\partial y)_x = 3y^2$. These can be confirmed with eqn (10.1)

$$\left(\frac{\partial z}{\partial x}\right)_y = \lim_{\delta x \to 0} \frac{3(x + \delta x)^2 + y^3 - (3x^2 + y^3)}{\delta x} = \lim_{\delta x \to 0} 6x + 3\delta x$$

and so on.

If we were standing on the side of a hill in real life, we wouldn't generally think of the partial derivatives as a means of talking about the local slope. Rather, we would tend to point up in the direction of the steepest ascent. This natural inclination is captured by the *gradient vector*, $\nabla z$, defined by

$$\nabla z = \left(\frac{\partial z}{\partial x}, \frac{\partial z}{\partial y}\right) \tag{10.2}$$

where we have omitted the subscripts on the partial derivatives, as being implicitly given, for simplicity of notation. Thus the quantities discussed earlier in eqn (10.1) correspond to the $x$ and $y$ components of $\nabla z$, a vector whose direction indicates the path of steepest ascent (or the greatest increase in $z$) and whose magnitude gives the value of the slope along it. Since the gradient vector of eqn (10.2) is two-dimensional, it is best to think of it as being an arrow in the contour map; the properties of the latter, marking as it does the points of equal height (where there is no change in $z$), means that $\nabla z$ will be perpendicular to the contour lines.

Although we have only explicitly considered a function of two variables, the ideas generalize quite readily to multi-parameter problems. For example, if we were dealing with a function of three variables, $\Phi = f(x, y, z)$ say, then the gradient vector would be given by

$$\nabla \Phi = \left(\frac{\partial \Phi}{\partial x}, \frac{\partial \Phi}{\partial y}, \frac{\partial \Phi}{\partial z}\right) \tag{10.3}$$

where the relevant subscripts have again been omitted from the partial derivatives. Thus $\partial \Phi/\partial x$ should really read $(\partial \Phi/\partial x)_{yz}$, to give the derivative of $\Phi$ with respect to $x$ at constant $y$ and $z$; $\partial \Phi/\partial y$ should be $(\partial \Phi/\partial y)_{xz}$, and so on. The previous contour lines have to be replaced by three-dimensional surfaces upon which the value of $\Phi$ is fixed. The gradient vector at a point $(x, y, z)$ is normal to the local 'isosurface' with its direction indicating the path of fastest increase in $\Phi$; the magnitude of $\nabla \Phi$ gives the rate of greatest change.

If $\Phi = 3x^2 + y^3 \sin x + \ln z$

$$\left(\frac{\partial \Phi}{\partial x}\right)_{yz} = 6x + y^3 \cos x$$

$$\left(\frac{\partial \Phi}{\partial y}\right)_{xz} = 3y^2 \sin x$$

$$\left(\frac{\partial \Phi}{\partial z}\right)_{xy} = \frac{1}{z}$$

## 10.2  Second and higher derivatives

Just as for ordinary derivatives in section 4.2, second and higher-order partial derivatives can also be evaluated. Thus with $z = f(x, y)$, for example,

$$\frac{\partial^2 z}{\partial x^2} = \frac{\partial}{\partial x}\left(\frac{\partial z}{\partial x}\right) \qquad (10.4)$$

where $\partial/\partial x$ is the partial differential operator for the 'rate of change with respect to $x$, with $y$ held fixed'; similarly, $\partial^2 z/\partial y^2 = \partial/\partial y\,(\partial z/\partial y)$. There are two other second partial derivatives but, some formal technicalities aside, these mixed terms are always equal

$$\frac{\partial}{\partial x}\left(\frac{\partial z}{\partial y}\right) = \frac{\partial}{\partial y}\left(\frac{\partial z}{\partial x}\right) \qquad (10.5)$$

and are denoted simply by $\partial^2 z/\partial x\,\partial y$. The relationship of eqn (10.5) is very important, and forms the basis of the derivation of many useful results.

$$\frac{\partial^2 \Phi}{\partial x^2} = 6 - y^3 \sin x$$

$$\frac{\partial^2 \Phi}{\partial y^2} = 6y \sin x$$

$$\frac{\partial^2 \Phi}{\partial x\,\partial y} = 3y^2 \cos x$$

## 10.3 Increments and chain rules

If $z = f(x,y)$, how much does $z$ change when $x$ and $y$ are varied by a small amount? Since $|\nabla z|$ tells us the rate of change of $z$ in the direction of greatest increase, the small change $\delta z$ caused by a tiny move in the $x$-$y$ plane is given by the dot product of $\nabla z$ and the path vector $\delta \mathbf{r} = (\delta x, \delta y)$

$$\delta z \approx \underbrace{|\nabla z|\,|\delta \mathbf{r}|\cos\theta}_{\nabla z \cdot \delta \mathbf{r}}$$

$$\delta z \approx \left(\frac{\partial z}{\partial x}\right)_y \delta x + \left(\frac{\partial z}{\partial y}\right)_x \delta y \qquad (10.6)$$

where equality is attained in the limit $\delta x \to 0$ and $\delta y \to 0$. The general multivariate extension of eqn (10.6) is straightforward; for $\Phi = f(x,y,z)$

$$\delta\Phi \approx \left(\frac{\partial\Phi}{\partial x}\right)_{yz} \delta x + \left(\frac{\partial\Phi}{\partial y}\right)_{xz} \delta y + \left(\frac{\partial\Phi}{\partial z}\right)_{xy} \delta z \qquad (10.7)$$

and so on.

Incremental formulae such as eqns (10.6) and (10.7) are very useful because they lead naturally to many different connections between differential coefficients. If we divide eqn (10.6) by $\delta u$, for example, where the quantity $u$ could be $x$, $y$, $z$, or something related to them, then

$$\frac{\delta z}{\delta u} \approx \left(\frac{\partial z}{\partial x}\right)_y \frac{\delta x}{\delta u} + \left(\frac{\partial z}{\partial y}\right)_x \frac{\delta y}{\delta u}$$

Following eqn (4.1), taking the limit $\delta u \to 0$ yields

$$\frac{dz}{du} = \left(\frac{\partial z}{\partial x}\right)_y \frac{dx}{du} + \left(\frac{\partial z}{\partial y}\right)_x \frac{dy}{du} \qquad (10.8)$$

whereas the same procedure carried out at constant $v$ gives

$$\left(\frac{\partial z}{\partial u}\right)_v = \left(\frac{\partial z}{\partial x}\right)_y \left(\frac{\partial x}{\partial u}\right)_v + \left(\frac{\partial z}{\partial y}\right)_x \left(\frac{\partial y}{\partial u}\right)_v \qquad (10.9)$$

Perhaps the simplest illustration of eqn (10.9) is when $u = z$ and $v = y$

$$\underbrace{\left(\frac{\partial z}{\partial z}\right)_y}_{1} = \left(\frac{\partial z}{\partial x}\right)_y\left(\frac{\partial x}{\partial z}\right)_y + \underbrace{\left(\frac{\partial z}{\partial y}\right)_x\left(\frac{\partial y}{\partial z}\right)_y}_{0}$$

which leads to

$$\left(\frac{\partial z}{\partial x}\right)_y = \frac{1}{(\partial x/\partial z)_y} \tag{10.10}$$

Thus the reciprocity relationship for partial derivatives only holds as long as the same parameters are held fixed. Similarly, letting $u = x$ and $v = z$ leads to

$$\left(\frac{\partial z}{\partial y}\right)_x\left(\frac{\partial y}{\partial x}\right)_z\left(\frac{\partial x}{\partial z}\right)_y = -1 \tag{10.11}$$

where we have also made use of eqn (10.10).

Although both eqns (10.10) and (10.11) follow from eqn (10.9), we must not forget eqn (10.8). This is important because it links the *total derivatives*, with ordinary d's, to the partial ones (having curly d's). Thus if $z = \mathsf{f}(x, y)$, but $x$ and $y$ are themselves functions of $t$ (time, say), then eqn (10.8) can be used to evaluate $\mathrm{d}z/\mathrm{d}t$ (by putting $u = t$) when direct substitution to obtain $z = \mathsf{f}(t)$ is difficult.

The chain rule of eqn (10.9) is also useful for changing variables in expressions that involve partial differential coefficients. In the polar to Cartesian transformation, $x = r \cos\theta$ and $y = r \sin\theta$, for example, putting $u = r$ and $v = \theta$ in eqn (10.9) yields the relationship

$$\left(\frac{\partial x}{\partial r}\right)_\theta = \cos\theta, \quad \left(\frac{\partial y}{\partial r}\right)_\theta = \sin\theta$$

$$\left(\frac{\partial z}{\partial r}\right)_\theta = \cos\theta\left(\frac{\partial z}{\partial x}\right)_y + \sin\theta\left(\frac{\partial z}{\partial y}\right)_x \tag{10.12}$$

The substitution of $u = \theta$ and $v = r$ gives $(\partial z/\partial\theta)_r$ in terms of $(\partial z/\partial x)_y$ and $(\partial z/\partial y)_x$. We need to be careful when evaluating higher-order derivatives, and remember how they are defined

$$\left(\frac{\partial z}{\partial\theta}\right)_r = -r\sin\theta\left(\frac{\partial z}{\partial x}\right)_y$$
$$+ r\cos\theta\left(\frac{\partial z}{\partial y}\right)_x$$

$$\frac{\partial^2 z}{\partial r^2} = \frac{\partial}{\partial r_\theta}\left(\frac{\partial z}{\partial r}\right)_\theta = \cos\theta\frac{\partial}{\partial r_\theta}\left(\frac{\partial z}{\partial x}\right)_y + \sin\theta\frac{\partial}{\partial r_\theta}\left(\frac{\partial z}{\partial y}\right)_x$$

where we have implicitly used the product rule of differentiation twice on the right-hand side, although one of the terms was a constant (at fixed $\theta$) in both cases. The awkward contributions, $\partial/\partial r_\theta(\partial z/\partial x)_y$ and $\partial/\partial r_\theta(\partial z/\partial y)_x$, can be ascertained by recognizing, from eqn (10.12), that the differential operator $\partial/\partial r_\theta$ can be written as

$$\frac{\partial}{\partial r_\theta}\left(\frac{\partial z}{\partial x}\right)_y = \cos\theta\frac{\partial^2 z}{\partial x^2}$$

$$+ \sin\theta\frac{\partial^2 z}{\partial x\,\partial y}$$

$$\frac{\partial}{\partial r_\theta} = \cos\theta\frac{\partial}{\partial x_y} + \sin\theta\frac{\partial}{\partial y_x}$$

and knowing that it obeys all the usual rules of algebra, and applies to quantities on its immediate right.

Finally, we should note that an alternative to the use of the sort of chain rules discussed above is often provided by implicit differentiation. For example, the easiest way to obtain $(\partial y/\partial x)_z$ when $z^2 = x^3 y + \ln(y)$ is to apply the partial differential operator $\partial/\partial x_z$ to both sides of the equation

$$2z\underbrace{\left(\frac{\partial z}{\partial x}\right)_z}_{0} = x^3 \left(\frac{\partial y}{\partial x}\right)_z + 3x^2 \underbrace{\left(\frac{\partial x}{\partial x}\right)_z}_{1} y + \frac{1}{y}\left(\frac{\partial y}{\partial x}\right)_z$$

A little algebraic rearrangement shows that $(\partial y/\partial x)_z = -3x^2 y^2/(x^3 y + 1)$. This also follows from $(\partial z/\partial x)_y$, $(\partial z/\partial y)_x$ and eqns (10.10) and (10.11).

## 10.4 Taylor series

In chapter 6, we considered the Taylor series as a means of approximating the curve $y = f(x)$ locally by a low-order polynomial; this idea can be extended to functions of two or more variables. In fact, eqn (10.6) represents the first-order expansion for $z = f(x, y)$

$$f(x, y) = f(x_0, y_0) + (x - x_0)\frac{\partial f}{\partial x}\bigg|_{x_0, y_0} + (y - y_0)\frac{\partial f}{\partial y}\bigg|_{x_0, y_0} + \cdots$$

$$\delta z = f(x, y) - f(x_0, y_0)$$
$$\delta x = x - x_0, \quad \delta y = y - y_0$$

where the partial derivatives are evaluated at $(x_0, y_0)$. This corresponds to approximating a two-dimensional surface by a tilted flat plane about the point of interest. A better estimate of the function is given by the addition of the three terms involving the second derivatives of $f(x, y)$

$$\frac{1}{2}\left[(x - x_0)^2 \frac{\partial^2 f}{\partial x^2}\bigg|_{x_0, y_0} + 2(x - x_0)(y - y_0)\frac{\partial^2 f}{\partial x \partial y}\bigg|_{x_0, y_0} + (y - y_0)^2 \frac{\partial^2 f}{\partial y^2}\bigg|_{x_0, y_0}\right]$$

which introduces a quadratic contribution. Higher-order terms become increasingly complicated and are rarely used.

For functions of more than two variables, it is best to generalize the Taylor series by adopting a 'matrix-vector' notation

$$f(\mathbf{x}) = f(\mathbf{x}_0) + (\mathbf{x} - \mathbf{x}_0)^{\mathsf{T}} \nabla f(\mathbf{x}_0) + \tfrac{1}{2}(\mathbf{x} - \mathbf{x}_0)^{\mathsf{T}} \nabla\nabla f(\mathbf{x}_0)(\mathbf{x} - \mathbf{x}_0) + \cdots$$

where $\mathbf{x}$ is a column matrix with components $(x_1, x_2, x_3, \ldots, x_N)$, $\nabla f(\mathbf{x}_0)$ is the N-dimensional gradient vector evaluated at the point $\mathbf{x}_0$, and $\nabla\nabla f(\mathbf{x}_0)$ is an $N \times N$ matrix whose $ij^{\text{th}}$ element is given by the second partial derivative $\partial^2 f/\partial x_i \partial x_j$ (also calculated at $\mathbf{x}_0$).

$$\left[\nabla\nabla f(\mathbf{x}_0)\right]_{ij} = \frac{\partial^2 f}{\partial x_i \partial x_j}\bigg|_{\mathbf{x}_0}$$

## 10.5 Maxima and minima

In section 4.6, we discussed the topic of maxima and minima for the curve $y = f(x)$; let's extend the analysis to encompass functions of several variables. The central idea of eqn (4.14), that a stationary point is a place where there is no slope, generalizes very easily

$$\nabla f = \mathbf{0} \qquad (10.13)$$

the difference being that we now have to deal with a gradient vector rather than a single derivative. The only way in which a vector can be equal to zero is if all its components are nought. This leads to a set of N simultaneous equations: $\partial f / \partial x_i = 0$ for $i = 1, 2, 3, \ldots, N$. Explicitly for the case of two parameters, $z = f(x, y)$, this means that

$$\left(\frac{\partial z}{\partial x}\right)_y = 0 \quad \text{and} \quad \left(\frac{\partial z}{\partial y}\right)_x = 0 \qquad (10.14)$$

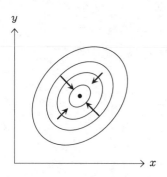

The best way to ensure that we find all the solutions, $(x, y)$, of eqn (10.14) is to factorize the partial derivatives as much as possible.

As a concrete example, consider the function $z = x^2 y^2 - x^2 - y^2$; the two conditions for its stationary points are

$$\left(\frac{\partial z}{\partial x}\right)_y = 2x\,(y^2 - 1) = 0 \quad \text{and} \quad \left(\frac{\partial z}{\partial y}\right)_x = 2y\,(x^2 - 1) = 0$$

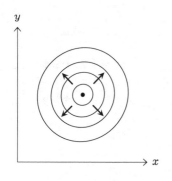

The partial derivative $(\partial z / \partial x)_y$ is zero if either $x = 0$ or $y = \pm 1$; $(\partial z / \partial y)_x$ is nought if either $y = 0$ or $x = \pm 1$. The simultaneous equations are satisfied at five points, therefore: $(0, 0)$, $(1, 1)$, $(1, -1)$, $(-1, 1)$, and $(-1, -1)$.

In the immediate neighbourhood of a stationary point, with $\nabla f(\mathbf{x_0}) = 0$, eqn (10.14) tells us that the behaviour of $f(\mathbf{x})$ is governed by the quadratic form of the $\nabla\nabla f$ matrix evaluated at $\mathbf{x_0}$

$$f(\mathbf{x}) = f(\mathbf{x_0}) + \tfrac{1}{2}(\mathbf{x} - \mathbf{x_0})^{\mathsf{T}} \nabla\nabla f(\mathbf{x_0})(\mathbf{x} - \mathbf{x_0})$$

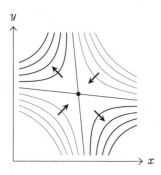

Following the discussion in section 9.8, the variation of $f(\mathbf{x})$ is parabolic along each of the eigenvector directions

$$f(\mathbf{x_0} + s\hat{\mathbf{x}}_i) = f(\mathbf{x_0}) + \lambda_i s^2$$

where $\lambda_i$ and $\hat{\mathbf{x}}_i$ are the corresponding $i^{\text{th}}$ eigenvalue and normalized eigenvector of $\nabla\nabla f(\mathbf{x_0})$ respectively. Thus $\mathbf{x_0}$ is a maximum if all the $\lambda_i$ are negative and a minimum if they're entirely positive; eigenvalues of mixed signs indicate a *saddle point*, with the function increasing in some directions and decreasing in others.

For the two-parameter case $z = f(x, y)$, the eigenvalues of the $2 \times 2$ matrix $\nabla\nabla z$ do not need to be calculated explicitly. This is because we can make use of the fact that the product of the eigenvalues, $\lambda_1 \lambda_2$, is equal to the determinant of the $\nabla\nabla z$ matrix

$$\nabla\nabla z = \begin{pmatrix} \dfrac{\partial^2 z}{\partial x^2} & \dfrac{\partial^2 z}{\partial x\,\partial y} \\[2mm] \dfrac{\partial^2 z}{\partial x\,\partial y} & \dfrac{\partial^2 z}{\partial y^2} \end{pmatrix}$$

$$\det\left(\nabla\nabla z\right) = \left(\frac{\partial^2 z}{\partial x^2}\right)\left(\frac{\partial^2 z}{\partial y^2}\right) - \left(\frac{\partial^2 z}{\partial x\,\partial y}\right)^2 \qquad (10.15)$$

If $\det(\nabla\nabla z) > 0$ both eigenvalues must have the same sign and the stationary point is either a maximum or a minimum; a negative value indicates a saddle point. A null result renders the test inconclusive, like $d^2y/dx^2 = 0$ in section 4.6. Having ascertained that $\det(\nabla\nabla z) > 0$, we need simply to look at the sign of $\partial^2 z/\partial x^2$ or $\partial^2 z/\partial y^2$, or their sum, to distinguish between a maximum and a minimum; it's negative for the former and positive for the latter.

$$\text{trace}(\nabla\nabla z) = \underbrace{\frac{\partial^2 z}{\partial x^2} + \frac{\partial^2 z}{\partial y^2}}_{\nabla^2 z}$$

For our earlier example $\partial^2 z/\partial x^2 = 2(y^2-1)$, $\partial^2 z/\partial y^2 = 2(x^2-1)$, and $\partial^2 z/\partial x\,\partial y = 4\,x\,y$. This leads to the identification of the stationary points at $\pm(1,1)$ and $\pm(1,-1)$ as saddle points because $\det(\nabla\nabla z) = -16 < 0$ for all four of them. With $\det(\nabla\nabla z) = 4 > 0$ and $\text{trace}(\nabla\nabla z) = -4 < 0$, the stationary point at $(0,0)$ is a maximum.

## 10.6 Constrained optimization

Often we are not interested in simply finding the maximum or minimum of a function, but are obliged to do so subject to certain constraints. A scenario from everyday life that illustrates this point is the task of trying to buy the best computer, car, or whatever, given a limited budget. In thermodynamics, we could be trying to ascertain the populations of energy levels subject to a fixed number of particles and the total energy of the system.

To explain the analysis, suppose we wish to find the 'optimum' value of the two-parameter function $z = f(x,y)$ subject to the condition $g(x,y) = 0$. For example, the maximum or minimum of $z = x^2 - x + 2y^2$ with $x^2 + y^2 = 1$. In this case we can easily put $y^2 = 1 - x^2$ from the constraint into $f(x,y)$ and reduce $z$ to a function of one variable: $z = 2 - x - x^2$. The required stationary points are then given by $dz/dx = -1 - 2x = 0$. The points $\left(-1/2, \pm\sqrt{3}/2\right)$, where $z = 9/4$, are conditional maxima as $d^2z/dx^2 = -2 < 0$.

The procedure outlined above is of limited utility because often $y$ cannot be expressed explicitly in terms of $x$, or the other way round, given $g(x,y) = 0$. A further issue is that some solutions can be missed unless we are careful and also check the cases when $dz/dy = 0$. An alternative approach for dealing with constrained optimization is provided by the method of *Lagrange undetermined multipliers*. This states that our problem is equivalent to finding the ordinary (unconditional) stationary points of a new function, $F(x,y)$, constructed from $f(x,y)$ and $g(x,y)$ according to

$$F(x,y) = f(x,y) + \lambda\, g(x,y) \qquad (10.16)$$

where $\lambda$ is a constant, the Lagrange multiplier, whose value is as yet unknown. While the rationale for eqn (10.16) is not easy to explain, its application is straightforward. The relevant values of $x$, $y$, and $\lambda$ can be obtained from

$$\left(\frac{\partial F}{\partial x}\right)_y = 0, \quad \left(\frac{\partial F}{\partial y}\right)_x = 0, \quad \text{and} \quad g(x,y) = 0$$

For our earlier example, these are: $2x(1+\lambda) - 1 = 0$, $2y(2+\lambda) = 0$, and $x^2 + y^2 - 1 = 0$ respectively. After some effort the stationary points are found to be at $\left(-1/2, \pm\sqrt{3}/2\right)$, as before, with $\lambda = -2$, but also $(\pm 1, 0)$, which are conditional minima with $\lambda = -1 \pm 1/2$.

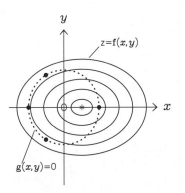

Although we have illustrated the use of Lagrange multipliers for the simplest setup, the method generalizes quite naturally to cases where there are several constraints or many variables in the function. Suppose we want to find the stationary points of $f(\mathbf{x})$, where the vector $\mathbf{x}$ has components $(x_1, x_2, \ldots, x_N)$, subject to $g_1(\mathbf{x}) = 0$, $g_2(\mathbf{x}) = 0$, $\ldots$, $g_M(\mathbf{x}) = 0$; then, we begin by constructing the hybrid function $F(\mathbf{x})$

$$F(\mathbf{x}) = f(\mathbf{x}) + \lambda_1 g_1(\mathbf{x}) + \lambda_2 g_2(\mathbf{x}) + \cdots + \lambda_M g_M(\mathbf{x})$$

where the $\lambda_i$, for $i = 1$ to M, are Lagrange multipliers. The desired values are given by the N+M parameters $x_1, x_2, \ldots, x_N, \lambda_1, \lambda_2, \ldots, \lambda_M$ which satisfy the N+M simultaneous equations $\partial F/\partial x_1 = 0$, $\partial F/\partial x_2 = 0$, $\ldots$, $\partial F/\partial x_N = 0$, $g_1(\mathbf{x}) = 0$, $g_2(\mathbf{x}) = 0$, $\ldots$, $g_M(\mathbf{x}) = 0$.

## 10.7 Worked examples

(1) Determine from first principles the partial derivatives $(\partial z/\partial x)_y$ and $(\partial z/\partial y)_x$ where $z = x^3/(1-y)$. Evaluate (by any means) $\partial^2 z/\partial x^2$ and $\partial^2 z/\partial y^2$, and check that $\partial^2 z/\partial x \partial y = \partial^2 z/\partial y \partial x$.

$$\left(\frac{\partial z}{\partial x}\right)_y = \lim_{\delta x \to 0} \frac{(x+\delta x)^3/(1-y) - x^3/(1-y)}{\delta x}$$

$$= \lim_{\delta x \to 0} \frac{x^3 + 3x^2 \delta x + 3x \delta x^2 + \delta x^3 - x^3}{(1-y)\delta x}$$

$$= \lim_{\delta x \to 0} \frac{3x^2 + 3x \delta x + \delta x^2}{1-y} = \underline{\frac{3x^2}{1-y}}$$

$$\left(\frac{\partial z}{\partial y}\right)_x = \lim_{\delta y \to 0} \frac{x^3/\left(1-[y+\delta y]\right) - x^3/(1-y)}{\delta y}$$

$$= \lim_{\delta y \to 0} \frac{x^3}{\delta y}\left[\frac{1}{(1-y-\delta y)} - \frac{1}{(1-y)}\right]$$

$$= \lim_{\delta y \to 0} \frac{x^3\left[(1-y) - (1-y-\delta y)\right]}{\delta y(1-y-\delta y)(1-y)}$$

$$= \lim_{\delta y \to 0} \frac{x^3}{(1-y-\delta y)(1-y)} = \underline{\frac{x^3}{(1-y)^2}}$$

$$\frac{\partial^2 z}{\partial x^2} = \frac{\partial}{\partial x_y}\left(\frac{\partial z}{\partial x}\right)_y = \frac{\partial}{\partial x_y}\left[\frac{3x^2}{1-y}\right] = \underline{\frac{6x}{1-y}}$$

$$\frac{\partial^2 z}{\partial y^2} = \frac{\partial}{\partial y_x}\left(\frac{\partial z}{\partial y}\right)_x = \frac{\partial}{\partial y_x}\left[\frac{x^3}{(1-y)^2}\right] = \frac{2\,x^3}{(1-y)^3}$$

$$\frac{\partial^2 z}{\partial x\,\partial y} = \frac{\partial}{\partial x_y}\left(\frac{\partial z}{\partial y}\right)_x = \frac{\partial}{\partial x_y}\left[\frac{x^3}{(1-y)^2}\right] = \frac{3\,x^2}{(1-y)^2}$$

$$\frac{\partial^2 z}{\partial y\,\partial x} = \frac{\partial}{\partial y_x}\left(\frac{\partial z}{\partial x}\right)_y = \frac{\partial}{\partial y_x}\left[\frac{3\,x^2}{1-y}\right] = \frac{3\,x^2}{(1-y)^2} = \frac{\partial^2 z}{\partial x\,\partial y}$$

Although we know that the mixed second derivatives should be equal, it's worth making sure that they are as a simple check on our calculation.

---

**(2)** If $f(x, y, z) = \cos(x\,y\,z)$, evaluate $\partial^3 f/\partial x\,\partial y\,\partial z$.

$$\frac{\partial f}{\partial z} = \left(\frac{\partial f}{\partial z}\right)_{xy} = -x y \sin(x y z)$$

$$\therefore \quad \frac{\partial^2 f}{\partial y\,\partial z} = \frac{\partial}{\partial y_{xz}}\left(\frac{\partial f}{\partial z}\right) = -x^2 y z \cos(x y z) - x \sin(x y z)$$

$$\therefore \quad \frac{\partial^3 f}{\partial x\,\partial y\,\partial z} = \frac{\partial}{\partial x_{yz}}\left(\frac{\partial^2 f}{\partial y\,\partial z}\right) = x^2 y^2 z^2 \sin(x y z) - 2 x y z \cos(x y z)$$
$$- x y z \cos(x y z) - \sin(x y z)$$
$$= (x^2 y^2 z^2 - 1)\sin(x y z) - 3 x y z \cos(x y z)$$

---

**(3)** The pressure $P$, volume $V$, and temperature $T$ of a mole of a real (rather than ideal) gas can be modelled with the *Van der Waals* equation

$$\left(P + \frac{a}{V^2}\right)(V - b) = RT$$

where $a$, $b$, and $R$ are constants. Verify that

$$\left(\frac{\partial P}{\partial V}\right)_T \left(\frac{\partial V}{\partial T}\right)_P \left(\frac{\partial T}{\partial P}\right)_V = -1$$

---

Van der Waals equation can be rearranged into the form $P = P(V, T)$

$$P = \frac{RT}{V - b} - \frac{a}{V^2} \quad -(1)$$

from which it follows that

$$\left(\frac{\partial P}{\partial V}\right)_T = -\frac{RT}{(V-b)^2} + \frac{2a}{V^3} \quad \text{and} \quad \left(\frac{\partial P}{\partial T}\right)_V = \frac{R}{V-b}$$

We could use the result in eqn (10.10) to write $(\partial T/\partial P)_V$ as the reciprocal of $(\partial P/\partial T)_V$, or we can obtain it directly by differentiating eqn (1) implicitly with respect to $P$ while treating $V$ like a constant

$$\underbrace{\left(\frac{\partial P}{\partial P}\right)_V}_{1} = \frac{R}{V-b}\left(\frac{\partial T}{\partial P}\right)_V + 0$$

Similarly, the implicit differentiation of eqn (1) with respect to $T$ at constant $P$ yields

$$\underbrace{\left(\frac{\partial P}{\partial T}\right)_P}_{0} = \frac{R}{V-b}\underbrace{\left(\frac{\partial T}{\partial T}\right)_P}_{1} - \frac{RT}{(V-b)^2}\left(\frac{\partial V}{\partial T}\right)_P + \frac{2a}{V^3}\left(\frac{\partial V}{\partial T}\right)_P$$

and simplifies to

$$-\frac{R}{V-b} = \left[-\frac{RT}{(V-b)^2} + \frac{2a}{V^3}\right]\left(\frac{\partial V}{\partial T}\right)_P$$

A comparison with the earlier partial derivative expressions shows this to be

$$-\left[\left(\frac{\partial T}{\partial P}\right)_V\right]^{-1} = \left(\frac{\partial P}{\partial V}\right)_T\left(\frac{\partial V}{\partial T}\right)_P$$

from which the desired expression follows readily.

---

**(4)** Show that if $f(u,v) = 0$ where $u = x+y$ and $v = x^2 + xy + z^2$ then $x+y = 2z\left[(\partial z/\partial y)_x - (\partial z/\partial x)_y\right]$.

---

$$f = f(u,v) = 0 \implies \delta f = 0 \approx \left(\frac{\partial f}{\partial u}\right)_v \delta u + \left(\frac{\partial f}{\partial v}\right)_u \delta v$$

$$\therefore \quad 0 = \left(\frac{\partial f}{\partial u}\right)_v\left(\frac{\partial u}{\partial x}\right)_y + \left(\frac{\partial f}{\partial v}\right)_u\left(\frac{\partial v}{\partial x}\right)_y$$

$$\text{and} \quad 0 = \left(\frac{\partial f}{\partial u}\right)_v\left(\frac{\partial u}{\partial y}\right)_x + \left(\frac{\partial f}{\partial v}\right)_u\left(\frac{\partial v}{\partial y}\right)_x$$

$$\left(\frac{\partial u}{\partial x}\right)_y = 1$$

i.e. $\quad \left(\frac{\partial f}{\partial u}\right)_v = -\left(\frac{\partial f}{\partial v}\right)_u\left[2x + y + 2z\left(\frac{\partial z}{\partial x}\right)_y\right] \quad$ — (1)

$$\left(\frac{\partial u}{\partial y}\right)_x = 1$$

and $\quad \left(\frac{\partial f}{\partial u}\right)_v = -\left(\frac{\partial f}{\partial v}\right)_u\left[x + 2z\left(\frac{\partial z}{\partial y}\right)_x\right] \quad$ — (2)

$$(1) \div (2) \Rightarrow \quad 2x + y + 2z\left(\frac{\partial z}{\partial x}\right)_y = x + 2z\left(\frac{\partial z}{\partial y}\right)_x$$

$$\text{i.e.} \quad x + y = 2z\left[\left(\frac{\partial z}{\partial y}\right)_x - \left(\frac{\partial z}{\partial x}\right)_y\right]$$

---

**(5)** Use the substitution $u = x + ct$ and $v = x - ct$, where $c$ is a constant, to reduce the one-dimensional wave equation $c^2\,\partial^2 z/\partial x^2 = \partial^2 z/\partial t^2$ to the form $\partial^2 z/\partial u\,\partial v = 0$.

---

$$z = z(u,v) \Rightarrow \quad \delta z \approx \left(\frac{\partial z}{\partial u}\right)_v \delta u + \left(\frac{\partial z}{\partial v}\right)_u \delta v$$

$$\therefore \quad \left(\frac{\partial z}{\partial x}\right)_t = \left(\frac{\partial z}{\partial u}\right)_v \underbrace{\left(\frac{\partial u}{\partial x}\right)_t}_{1} + \left(\frac{\partial z}{\partial v}\right)_u \underbrace{\left(\frac{\partial v}{\partial x}\right)_t}_{1} = \left(\frac{\partial z}{\partial u}\right)_v + \left(\frac{\partial z}{\partial v}\right)_u$$

$$\text{and} \quad \left(\frac{\partial z}{\partial t}\right)_x = \left(\frac{\partial z}{\partial u}\right)_v \underbrace{\left(\frac{\partial u}{\partial t}\right)_x}_{c} + \left(\frac{\partial z}{\partial v}\right)_u \underbrace{\left(\frac{\partial v}{\partial t}\right)_x}_{-c} = c\left(\frac{\partial z}{\partial u}\right)_v - c\left(\frac{\partial z}{\partial v}\right)_u$$

$$\frac{\partial^2 z}{\partial x^2} = \frac{\partial}{\partial x_t}\left(\frac{\partial z}{\partial x}\right)_t = \left[\frac{\partial}{\partial u_v} + \frac{\partial}{\partial v_u}\right]\left[\left(\frac{\partial z}{\partial u}\right)_v + \left(\frac{\partial z}{\partial v}\right)_u\right]$$

$$= \frac{\partial}{\partial u_v}\left(\frac{\partial z}{\partial u}\right)_v + \frac{\partial}{\partial u_v}\left(\frac{\partial z}{\partial v}\right)_u + \frac{\partial}{\partial v_u}\left(\frac{\partial z}{\partial u}\right)_v + \frac{\partial}{\partial v_u}\left(\frac{\partial z}{\partial v}\right)_u$$

$$= \frac{\partial^2 z}{\partial u^2} + 2\frac{\partial^2 z}{\partial u\,\partial v} + \frac{\partial^2 z}{\partial v^2} \qquad - (1)$$

$$\frac{\partial^2 z}{\partial t^2} = \frac{\partial}{\partial t_x}\left(\frac{\partial z}{\partial t}\right)_x = \left[c\frac{\partial}{\partial u_v} - c\frac{\partial}{\partial v_u}\right]\left[c\left(\frac{\partial z}{\partial u}\right)_v - c\left(\frac{\partial z}{\partial v}\right)_u\right]$$

$$= c^2\frac{\partial^2 z}{\partial u^2} - 2c^2\frac{\partial^2 z}{\partial u\,\partial v} + c^2\frac{\partial^2 z}{\partial v^2} \qquad - (2)$$

$$(1) - \frac{(2)}{c^2} \Rightarrow \quad \frac{\partial^2 z}{\partial x^2} - \frac{1}{c^2}\frac{\partial^2 z}{\partial t^2} = 4\frac{\partial^2 z}{\partial u\,\partial v}$$

$$\text{But} \quad \frac{\partial^2 z}{\partial x^2} - \frac{1}{c^2}\frac{\partial^2 z}{\partial t^2} = 0 \quad \therefore \quad \frac{\partial^2 z}{\partial u\,\partial v} = 0$$

The partial differential operators, $\partial/\partial x_t$ and $\partial/\partial t_x$, are easily ascertained from the expressions for the first derivatives, $(\partial z/\partial x)_t$ and $(\partial z/\partial t)_x$, as long as we are careful to rearrange the equations so that $z$ always appears on the far right. For example,

$$\left(\frac{\partial z}{\partial t}\right)_x = \frac{\partial}{\partial t_x}(z) = \left[c\frac{\partial}{\partial u_v} - c\frac{\partial}{\partial v_u}\right](z) = c\left(\frac{\partial z}{\partial u}\right)_v - c\left(\frac{\partial z}{\partial v}\right)_u$$

$$\therefore \quad \frac{\partial}{\partial t_x} = c\frac{\partial}{\partial u_v} - c\frac{\partial}{\partial v_u}$$

Differential operators obey the same rules as ordinary algebraic expressions, such as for multiplying out brackets and so on, and apply to everything on their immediate right. The wave equation case was easy because $c$ was a constant; if its equivalent had been a function of $x$ and $t$, and therefore (implicitly) one of $u$ and $v$, say $\beta(u,v)$, then we would have needed to use the product rule several times, as in

$$\beta(u,v)\frac{\partial}{\partial u_v}\left[\beta(u,v)\left(\frac{\partial z}{\partial u}\right)_v\right] = \beta(u,v)\left[\beta(u,v)\frac{\partial^2 z}{\partial u^2} + \left(\frac{\partial\beta}{\partial u}\right)_v\left(\frac{\partial z}{\partial u}\right)_v\right]$$

so that the related algebra (but not the principle of the procedure) can become rather messy.

---

**(6)** Find and classify all the (real) stationary values of the function
$f(x,y) = y^2(a^2+x^2) - x^2(2a^2-x^2)$, where $a$ is a (real) constant.

---

$$\left(\frac{\partial f}{\partial x}\right)_y = 2xy^2 - 4a^2x + 4x^3 = 2x(y^2 - 2a^2 + 2x^2)$$

$$\left(\frac{\partial f}{\partial y}\right)_x = 2y(a^2+x^2)$$

$$\therefore \quad \frac{\partial^2 f}{\partial x^2} = 2y^2 - 4a^2 + 12x^2, \quad \frac{\partial^2 f}{\partial y^2} = 2(a^2+x^2), \quad \frac{\partial^2 f}{\partial x\partial y} = 4xy$$

For stationary points, $\left(\dfrac{\partial f}{\partial x}\right)_y = 0$ and $\left(\dfrac{\partial f}{\partial y}\right)_x = 0$

$$\left.\begin{array}{ll}\therefore & x(y^2 - 2a^2 + 2x^2) = 0 \\ \text{and} & y(a^2+x^2) = 0\end{array}\right\} \quad \begin{array}{l} x = 0 \quad \text{and} \quad y = 0 \\ \text{or} \quad y = 0 \quad \text{and} \quad x = \pm a\end{array}$$

To classify stationary points, we first need to consider sign of

$$\det(\nabla\nabla f) = \left(\frac{\partial^2 f}{\partial x^2}\right)\left(\frac{\partial^2 f}{\partial y^2}\right) - \left(\frac{\partial^2 f}{\partial x\partial y}\right)^2$$

$\left[\text{real } a \implies a^2 \geqslant 0\right]$

At $(0,0)$, $\det(\nabla\nabla f) = -8a^4 < 0$ $\quad\therefore\quad$ $(0,0)$ is a saddle point

At $(\pm a, 0)$, $\det(\nabla\nabla f) = 32a^4 > 0$ $\quad\therefore\quad$ maximum or minimum

Now need sign of $\nabla^2 f = \dfrac{\partial^2 f}{\partial x^2} + \dfrac{\partial^2 f}{\partial y^2}$

At $(\pm a, 0)$,   $\nabla^2 f = 12 a^2 > 0$   $\therefore$  $(\pm a, 0)$ are both minima

When trying to find the stationary points, it is worth factorizing the first derivatives as much as possible first. In this example, $(\partial f/\partial x)_y = 0$ tells us that either $x = 0$ or $y^2 - 2a^2 + 2x^2 = 0$ and $(\partial f/\partial y)_x = 0$ means that either $y = 0$ or $a^2 + x^2 = 0$. By considering each of the four combinations which make both the first derivatives simultaneously equal to zero, we are assured of finding all the stationary points.

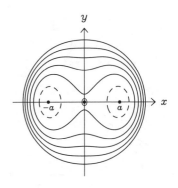

---

**(7)** Use the method of Lagrange multipliers to find the stationary values of the function $e^{-xy}$ subject to the condition $x^2 + y^2 = 1$.

---

For stationary points of $f(x, y) = e^{-xy}$ subject to $g(x, y) = x^2 + y^2 - 1 = 0$

let $F(x, y) = f(x, y) + \lambda g(x, y) = e^{-xy} + \lambda(x^2 + y^2 - 1)$

Then   $\left(\dfrac{\partial F}{\partial x}\right)_y = 0$ $\Rightarrow$   $y e^{-xy} = 2x\lambda$   — (1)

and   $\left(\dfrac{\partial F}{\partial y}\right)_x = 0$ $\Rightarrow$   $x e^{-xy} = 2y\lambda$   — (2)

and   $g(x, y) = 0$ $\Rightarrow$   $x^2 + y^2 = 1$   — (3)

$\dfrac{(1)}{(2)}$ $\Rightarrow$ $\dfrac{y}{x} = \dfrac{x}{y}$   $\therefore$ $y^2 = x^2$   i.e. $y = \pm x$

$\therefore$ (3) $\Rightarrow$   $2x^2 = 1$   $\therefore$ $x = \pm\dfrac{1}{\sqrt{2}}$

$\therefore$ Stationary points are at $\pm\left(\dfrac{1}{\sqrt{2}}, \dfrac{1}{\sqrt{2}}\right)$,  with $f = e^{-1/2}$

and at $\pm\left(\dfrac{1}{\sqrt{2}}, \dfrac{-1}{\sqrt{2}}\right)$,  with $f = e^{1/2}$

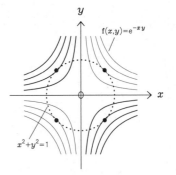

---

**(8)** The 'best' straight line through a set of N noisy data-points, $(X_i, Y_i)$ for $i = 1$ to N, is often taken to be the one that minimizes the function

$$Q(m, c) = \sum_{i=1}^{N} \left[m X_i + c - Y_i\right]^2$$

where $m$ and $c$ are the gradient and intercept of the line, respectively. Derive the formulae for $m$ and $c$ in terms of the $X_i$ and $Y_i$.

For minimum Q, $\quad \left(\dfrac{\partial Q}{\partial m}\right)_c = 0 \quad$ and $\quad \left(\dfrac{\partial Q}{\partial c}\right)_m = 0 \quad$ — (1)

Using the linearity of the differential operator, and the chain rule,

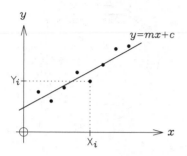

$$\left(\frac{\partial Q}{\partial m}\right)_c = \sum_{i=1}^{N} 2\left[mX_i + c - Y_i\right]\underbrace{\frac{\partial}{\partial m_c}\left[mX_i + c - Y_i\right]}_{X_i}$$

and $\quad \left(\dfrac{\partial Q}{\partial c}\right)_m = \displaystyle\sum_{i=1}^{N} 2\left[mX_i + c - Y_i\right]\underbrace{\frac{\partial}{\partial c_m}\left[mX_i + c - Y_i\right]}_{1}$

Expanding the summations into three separate terms, and taking the factors that do not depend on the index $i$ outside the $\Sigma$-symbols,

$$(1) \Rightarrow \quad m\underbrace{\sum_{i=1}^{N} X_i^2}_{\alpha} + c\underbrace{\sum_{i=1}^{N} X_i}_{\beta} = \underbrace{\sum_{i=1}^{N} X_i Y_i}_{\lambda}$$

$$\begin{pmatrix} \alpha & \beta \\ \beta & N \end{pmatrix}\begin{pmatrix} m \\ c \end{pmatrix} = \begin{pmatrix} \lambda \\ \mu \end{pmatrix}$$

and $\quad m\underbrace{\sum_{i=1}^{N} X_i}_{\beta} + c\underbrace{\sum_{i=1}^{N} 1}_{N} = \underbrace{\sum_{i=1}^{N} Y_i}_{\mu}$

Hence the required gradient and intercept satisfy two linear simultaneous equations, with coefficients $\alpha$, $\beta$, $\lambda$, and $\mu$ related to the data-points as indicated above, and have the unique solution

$$m = \frac{N\lambda - \beta\mu}{\alpha N - \beta^2} \quad \text{and} \quad c = \frac{\alpha\mu - \beta\lambda}{\alpha N - \beta^2}$$

For the sole stationary point of the quadratic function $Q(m,c)$ to be a minimum rather than a saddle point we require that, in addition to eqn (1), $\det(\nabla\nabla Q) = \alpha N - \beta^2 > 0$.

# 11 Line integrals

## 11.1 Line integrals

In this chapter, we will be considering the topic of *line integrals*. As a simple example of what they are and how they arise, let's think about an elementary formula met in school physics: 'Work done $=$ force $\times$ distance', or $W = F\,L$. Since both the force $\mathbf{F}$ and the related displacement $\mathbf{L}$ are actually vectors, the multiplication should really be a dot product: $W = \mathbf{F} \cdot \mathbf{L}$. If $\mathbf{F}$ varies with position $\mathbf{r}$, or the motion is not in a straight line, $W$ has to be calculated by adding up the small contributions of work done, $\delta W \approx \mathbf{F}(\mathbf{r}) \cdot \delta \mathbf{L}$, from tiny step-lengths $\delta \mathbf{L}$ all along the path travelled. The sum is exact in the limit $\delta \mathbf{L} \to 0$

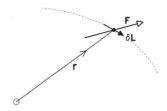

$$W = \lim_{\delta \mathbf{L} \to 0} \sum_{\text{path}} \delta W = \int_{\text{path}} dW = \int_{\text{path}} \mathbf{F}(\mathbf{r}) \cdot d\mathbf{L} \qquad (11.1)$$

Since the relevant position vectors lie along the path, $d\mathbf{L} = d\mathbf{r}$.

If the motion is in a two-dimensional plane then, in Cartesian coordinates, $\mathbf{r} = (x, y)$, $d\mathbf{r} = (dx, dy)$, and $\mathbf{F} = (F_x, F_y)$. Hence

$$W = \int_{\text{path}} \mathbf{F} \cdot d\mathbf{r} = \int_{\text{path}} F_x\, dx + F_y\, dy \qquad (11.2)$$

This can be written as the sum of two separate integrals, for $dx$ and $dy$, and the components of the force, $F_x$ and $F_y$, are functions of both $x$ and $y$ in general. The path will typically go from some point A, with coordinates $(x_A, y_A)$, to B, at $(x_B, y_B)$, along a trajectory defined by the curve $y = f(x)$. This relationship between $x$ and $y$ enables us to express eqn (11.2) entirely in terms of $x$

$$\int_{\text{path}} F_x(x,y)\, dx + F_y(x,y)\, dy = \int_{x_A}^{x_B} \left[ F_x(x, f(x)) + F_y(x, f(x))\, f'(x) \right] dx$$

$$dy = \underbrace{\frac{dy}{dx}}_{f'(x)}\, dx$$

We could, of course, have converted everything into $y$ instead, if that was more convenient.

As a concrete illustration of the discussion above, let us calculate the integral of $y^3\, dx + x\, dy$ from $(0,0)$ to $(1,1)$ along two different paths: (a) $y = x^2$; (b) the straight lines from $(0,0)$ to $(1,0)$, and from $(1,0)$ to $(1,1)$. For the first

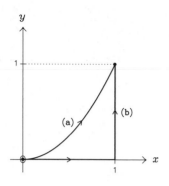

case, the substitution of $y = x^2$ and $dy = 2x\,dx$ leads to

$$\int_{path\,(a)} y^3\,dx + x\,dy = \int_0^1 (x^6 + 2x^2)\,dx = \left[\frac{x^7}{7} + \frac{2x^3}{3}\right]_0^1 = \frac{17}{21}$$

For the two-part route we have $y = 0$ and $dy = 0$ between $(0,0)$ and $(1,0)$, and $x = 1$ and $dx = 0$ on going from $(1,0)$ to $(1,1)$; hence, the sum of these straight line contributions gives

$$\int_{path\,(b)} y^3\,dx + x\,dy = \int_{x=0}^{x=1} 0\,dx + \int_{y=0}^{y=1} dy = 0 + [y]_0^1 = 1 \neq \frac{17}{21}$$

The result of a line integral is, in general, dependent on the path followed from start to finish. The exception to this is when the integrand is an *exact differential*, which we will come to shortly, and then the line integral is independent of the route taken, and a complicated path can be replaced by a simpler one to facilitate the calculation.

We should mention an alternative form of the line integral which is appropriate when dealing with scalar quantities

$$I = \int_{path} \Phi(x,y)\,dL \tag{11.3}$$

where the integration is with respect to the arc-length, $dL = |dL|$, along the curve $y = f(x)$; the path may also be specified by a parametric equation such as $x = g(t)$ and $y = h(t)$. The way to handle eqn (11.3) is very similar to eqn (11.2), except that $dL$ is related to $dx$ or $dt$ by

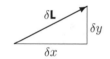

$$dL = \sqrt{1 + \left(\frac{dy}{dx}\right)^2}\,dx \quad \text{or} \quad dL = \sqrt{\left(\frac{dx}{dt}\right)^2 + \left(\frac{dy}{dt}\right)^2}\,dt \tag{11.4}$$

which essentially follows from Pythagoras' theorem, $dL^2 = dx^2 + dy^2$. If $y = x^2$, for example, $dL = \sqrt{1 + 4x^2}\,dx$. It almost goes without saying that $\Phi(x,y)$ can be written as $\Phi(x, f(x))$ or $\Phi(g(t), h(t))$, depending on the format used for the integration path. For the case when $\Phi(x,y) = 1$, eqn (11.3) returns the length of the path.

## 11.2 Exact differentials

Sometimes we are faced with expressions of the type $P(x,y)\,dx + Q(x,y)\,dy$, where P and Q are arbitrary functions of $x$ and $y$, and a question that arises is whether this constitutes the formula for the increment, $df$, of some quantity $f(x,y)$. If it does, then $P\,dx + Q\,dy$ is said to be an exact differential; otherwise, it's not. So, how do we test for it?

If such a function $f(x,y)$ existed, then a comparison with the limiting form of eqn (10.6) shows that

$$df = \left(\frac{\partial f}{\partial x}\right)_y dx + \left(\frac{\partial f}{\partial y}\right)_x dy$$

$$P(x,y) = \left(\frac{\partial f}{\partial x}\right)_y \quad \text{and} \quad Q(x,y) = \left(\frac{\partial f}{\partial y}\right)_x \tag{11.5}$$

While this must be true, it's not very helpful as it stands because it requires us to know $f(x,y)$ explicitly. A useful test for exactness follows, however, when it is combined with eqn (10.5)

$$\left(\frac{\partial Q}{\partial x}\right)_y = \left(\frac{\partial P}{\partial y}\right)_x \tag{11.6}$$

Thus, for example, $x\,y^2\,\mathrm{d}x + x^2\,y\,\mathrm{d}y$ is an exact differential but $y^3\,\mathrm{d}x + x\,\mathrm{d}y$ is not.

But why should we be interested in exact differentials at all? To understand that, we need to consider the integral of $P\,\mathrm{d}x + Q\,\mathrm{d}y$ from some initial point A, with coordinates $(x_A, y_A)$, to B, defined by $(x_B, y_B)$. If $P\,\mathrm{d}x + Q\,\mathrm{d}y = \mathrm{d}f$, then

$$\int_A^B P\,\mathrm{d}x + Q\,\mathrm{d}y = \int_A^B \mathrm{d}f = \Big[f(x,y)\Big]_A^B = f(x_B,y_B) - f(x_A,y_A)$$

In other words, the integral depends only on the end-points, A and B, and not on the path taken for going from one to the other. If $P\,\mathrm{d}x + Q\,\mathrm{d}y$ is not an exact differential, the first step above does not follow; consequently, we must specify the details of the integration path used and deal with any ensuing complications in the calculation.

The physical relevance of the discussion in this section is that exact differentials are related to *conservative fields* (or forces) and *state functions*. For example, the change in the gravitational potential energy of an object depends only on the difference in its height and not on any other characteristics of the motion. Similarly, in thermodynamics, the value of a function-of-state depends literally on the state (temperature, pressure, volume, etc.) of the system and not on how it got there.

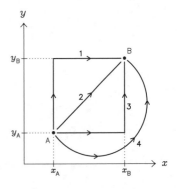

## 11.3 **Worked examples**

**(1)** Show that the integral of $y^3\,\mathrm{d}x + 3x\,y^2\,\mathrm{d}y$ is independent of path by evaluating it over the following two routes between the origin and the point $(1,1)$: (a) $y = x^2$; and (b) the straight lines from $(0,0)$ to $(1,0)$, and from $(1,0)$ to $(1,1)$.

(a)  Along $y = x^2$, $\quad \mathrm{d}y = 2x\,\mathrm{d}x$

$$\therefore \quad \int_{\text{path (a)}} y^3\,\mathrm{d}x + 3x\,y^2\,\mathrm{d}y = \int_{x=0}^{x=1} (x^6 + 6x^6)\,\mathrm{d}x = \Big[x^7\Big]_0^1 = \underline{1}$$

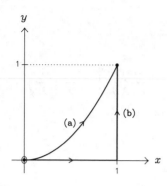

(b)   From  $(0,0)$ to $(1,0)$,   $y=0$  and  $dy=0$

$$\therefore \int_{(0,0)}^{(1,0)} y^3 \, dx + 3xy^2 \, dy = \int_{x=0}^{x=1} 0 \, dx = \left[K\right]_0^1 = 0$$

From  $(1,0)$ to $(1,1)$,   $x=1$  and  $dx=0$

$$\therefore \int_{(1,0)}^{(1,1)} y^3 \, dx + 3xy^2 \, dy = \int_{y=0}^{y=1} 3y^2 \, dy = \left[y^3\right]_0^1 = 1$$

$$\int_{\text{path (b)}} y^3 \, dx + 3xy^2 \, dy = \text{sum of the above} = 0 + 1 = \underline{1}$$

Although the evaluation of the integral of $y^3 \, dx + 3xy^2 \, dy$ over the two paths is a useful confirmation that it is independent of path, the formal proof comes from the satisfaction of the condition that

$$\frac{\partial}{\partial y_x}\left(y^3\right) = \frac{\partial}{\partial x_y}\left(3xy^2\right)$$

**(2)** Evaluate the integral of $xy \, dL$ along the two paths, (a) and (b), of the previous example.

$$\delta L = \sqrt{\delta x^2 + \delta y^2}$$

$$dL = \sqrt{1 + \left(\tfrac{dy}{dx}\right)^2} \, dx \qquad \text{(by Pythagoras)}$$

(a)   Along  $y = x^2$,    $\frac{dy}{dx} = 2x$

$$\therefore \int_{\text{path (a)}} xy \, dL = \int_{x=0}^{x=1} x^3 \sqrt{1 + 4x^2} \, dx$$

Put   $u^2 = 1 + 4x^2$    $\therefore 2u \, du = 8x \, dx$

$$\therefore \int_{\text{path (a)}} xy \, dL = \int_{u=1}^{u=\sqrt{5}} \underbrace{\frac{(u^2-1)}{4}}_{x^2} \, u \, \underbrace{\frac{u \, du}{4}}_{x \, dx} = \frac{1}{16} \int_1^{\sqrt{5}} \left(u^4 - u^2\right) du$$

$$= \frac{1}{16} \left[\frac{u^5}{5} - \frac{u^3}{3}\right]_1^{\sqrt{5}}$$

$$\therefore \int_{path\,(a)} xy\,dL = \frac{1}{16}\left[\frac{25\sqrt{5}}{5} - \frac{5\sqrt{5}}{3} - \frac{1}{5} + \frac{1}{3}\right]$$

$$= \frac{1}{16} \times \frac{(75-25)\sqrt{5} - 3 + 5}{15} = \frac{25\sqrt{5}+1}{120} \approx \frac{56.9}{120}$$

(b)   From $(0,0)$ to $(1,0)$,   $y = 0$ and $dL = dx$

$$\therefore \int_{(0,0)}^{(1,0)} xy\,dL = \int_{x=0}^{x=1} 0\,dx = \left[K\right]_0^1 = 0$$

From $(1,0)$ to $(1,1)$,   $x = 1$ and $dL = dy$

$$\therefore \int_{(1,0)}^{(1,1)} xy\,dL = \int_{y=0}^{y=1} y\,dy = \left[\frac{1}{2}y^2\right]_0^1 = \frac{1}{2}$$

$$\int_{path\,(b)} xy\,dL = \text{sum of the above} = 0 + \frac{1}{2} = \frac{1}{2}$$

---

**(3)** For one mole of an ideal gas, tiny changes in volume and temperature, $dV$ and $dT$ respectively, result in a corresponding heat change, $dq$, of $C_V\,dT + (RT/V)\,dV$, where R is the gas constant and the heat capacity, $C_V$, is independent of volume. Prove that this expression for $dq$ is not an exact differential, but that it becomes so when divided through by $T$. Comment on the relevance of this to thermodynamics.

$$\left.\begin{array}{l} \dfrac{\partial}{\partial V}_T(C_V) = 0 \\[2em] \dfrac{\partial}{\partial T}_V\left(\dfrac{RT}{V}\right) = \dfrac{R}{V} \end{array}\right\} \quad \therefore\ \dfrac{\partial}{\partial V}_T(C_V) \neq \dfrac{\partial}{\partial T}_V\left(\dfrac{RT}{V}\right)$$

$$\Rightarrow\ C_V\,dT + \frac{RT}{V}\,dV \ \text{is not an exact differential}$$

$$\left.\begin{array}{l} \dfrac{\partial}{\partial V}_T\left(\dfrac{C_V}{T}\right) = 0 \\[2em] \dfrac{\partial}{\partial T}_V\left(\dfrac{R}{V}\right) = 0 \end{array}\right\} \quad \therefore\ \dfrac{\partial}{\partial V}_T\left(\dfrac{C_V}{T}\right) = \dfrac{\partial}{\partial T}_V\left(\dfrac{R}{V}\right)$$

$$\Rightarrow\ \frac{C_V}{T}\,dT + \frac{R}{V}\,dV \ \text{is an exact differential}$$

*Heat* is not a state function because its incremental change, $dq$, at least for an ideal gas, is not an exact differential: $dq$ is not the increment of some function of volume and temperature, $q(V,T)$. The heat change between two states, $\int_1^2 dq$, defined by $(V_1, T_1)$ and $(V_2, T_2)$, depends not only on the difference in $V$ and $T$ but also on how the transition was made. Was the volume held fixed first and the temperature allowed to change, for example, or the other way around?

On dividing by the temperature, we have found that $dq/T$ is an exact differential. This means that it corresponds to a tiny change, $dS$, in a quantity which is the increment of some function of volume and temperature, $S(V,T)$. The latter is known as the *entropy* and is a state function. The change in entropy

$$\int_1^2 dS = \int_1^2 \frac{dq}{T} = \left[ S(V,T) \right]_1^2 = S(V_2, T_2) - S(V_1, T_1)$$

depends only on the initial and final state, and not on the details of the transition between them.

Incidentally, $1/T$ is said to be the *integrating factor* for this problem: a multiplicative 'weighting term' which converts an inexact differential into an exact one. We could have derived it by guessing that it was a function of temperature only, $w = w(T)$, from the requirement

$$\left( \frac{\partial w}{\partial T} \right)_V = \frac{dw}{dT}$$

$$\frac{\partial}{\partial V_T} \left[ C_V\, w(T) \right] = \frac{\partial}{\partial T_V} \left[ \frac{RT}{V}\, w(T) \right]$$

which leads to a simple first-order ordinary differential equation

$$0 = \frac{RT}{V} \frac{dw}{dT} + \frac{R}{V} w$$

that has the solution $w = A/T$ where $A$ is a constant.

# 12 Multiple integrals

## 12.1 Physical examples

In chapter 5, on integration, we were concerned with the area under the curve $y = f(x)$. Following our discussion about partial differentiation, however, we know that many quantities of interest depend on several variables. The topic of *multiple integrals* is, therefore, a natural extension of our earlier ideas to deal with multi-parameter problems.

To get a feel for multiple integrals, let's consider a couple of physical examples. Suppose we wish to calculate the force exerted on a wall by a gale. If the pressure $P$ was constant across the face with area $A$, then the total force is simply $P \times A$. With a varying pressure $P(x, y)$, the calculation has to be done by working out the force on tiny segments of area, $\delta \text{Force} = P(x, y)\, \delta x\, \delta y$, and adding up all the contributions over the entire wall. In the limit of $\delta x \to 0$ and $\delta y \to 0$,

$$\text{Force} = \iint_{\text{wall}} d\text{Force} = \iint_{\text{wall}} P(x, y)\, dx\, dy \qquad (12.1)$$

$$\delta\text{Area} = \delta x\, \delta y$$

where the *double integral* indicates that the infinitesimal summation is being carried out over a two-dimensional surface (in the $x$ and $y$ directions). Incidentally, if the wall does not have a conventional (rectangular) shape then its area can be be calculated in a similar manner

$$\text{Area} = \iint_{\text{wall}} d\text{Area} = \iint_{\text{wall}} dx\, dy \qquad (12.2)$$

The double integral is also called a *surface integral*.

Another illustration is provided by quantum mechanics where the modulus-squared of the wavefunction, $|\psi(x, y, z)|^2$, of an electron (say) gives the *probability density* of finding it at some point in space. The chances that the electron is in a small (cuboid) region of volume $\delta x\, \delta y\, \delta z$ is then $|\psi(x, y, z)|^2\, \delta x\, \delta y\, \delta z$. Hence, the probability of finding it within a finite domain $V$ is given by

$$\text{Probability} = \iiint_{V} |\psi(x, y, z)|^2\, dx\, dy\, dz \qquad (12.3)$$

$$d\text{Volume} = dx\, dy\, dz$$

which is known as a *triple*, or *volume*, *integral*.

## 12.2 The order of integration

As a concrete example of how to calculate multiple integrals, let's consider a very easy case; namely, working out the formula for the area of a right-angled triangle. The simplest setup is to place one apex at the origin and have the base, of length $L$, along the $x$-axis; the vertical edge, of height $H$, can then rise up from $(L,0)$ to $(L,H)$. If we take a set of small elementary boxes, of area $\delta x\,\delta y$, and stack them up parallel to the $y$-axis at some point along the $x$-axis (where $0 \leqslant x \leqslant L$), then we will obtain a narrow vertical strip going from $y=0$ to $y=Hx/L$. In the limit of $\delta x \to 0$ and $\delta y \to 0$, its area is given by

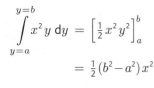

$$\text{Area of vertical strip} \;=\; \int_{y=0}^{y=\frac{Hx}{L}} \mathrm{d}\text{Area} \;=\; \mathrm{d}x \int_{y=0}^{y=\frac{Hx}{L}} \mathrm{d}y$$

where we have taken $\mathrm{d}x$ out of the $y$-integral because it does not depend on $y$. The area of the triangle is then the sum of all such contributions, with $x$ ranging from $x=0$ to $x=L$

$$\text{Area of triangle} \;=\; \int_{x=0}^{x=L} \mathrm{d}x \int_{y=0}^{y=\frac{Hx}{L}} \mathrm{d}y \tag{12.4}$$

where the convention is that the integral on the far right is evaluated first. Hence, the expected result emerges

$$\text{Area} \;=\; \int_0^L \Big[ y \Big]_0^{\frac{Hx}{L}} \mathrm{d}x \;=\; \int_0^L \frac{Hx}{L}\,\mathrm{d}x \;=\; \frac{H}{L} \Big[ \tfrac{1}{2} x^2 \Big]_0^L \;=\; \tfrac{1}{2} HL$$

$$\int_{y=a}^{y=b} x^2 y\,\mathrm{d}y \;=\; \Big[ \tfrac{1}{2} x^2 y^2 \Big]_a^b$$

$$= \tfrac{1}{2}\left(b^2 - a^2\right) x^2$$

While this may seem like a rather tortuous way of getting to an obvious answer, the method would become essential if the shape was more complicated or the integral was of the type in eqn (12.1) instead of eqn (12.2); that is to say, with $P(x,y)$ not uniform. In a manner similar to partial derivatives, any occurrence of $x$ inside the $y$-integral is treated like a constant (and vice versa).

We could, of course, have carried out the calculation by stacking the elementary boxes along the $x$-direction first, going from $x = L\,y/H$ to $x = L$, and then adding up the contributions from all the horizontal strips, so that the $y$-integral ranges from $y=0$ to $y=H$

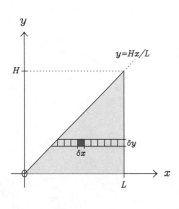

$$\text{Area of triangle} \;=\; \int_{y=0}^{y=H} \mathrm{d}y \int_{x=\frac{Ly}{H}}^{x=L} \mathrm{d}x \tag{12.5}$$

Upon evaluating the $x$-integral, followed by the one in $y$, we again obtain the expected result

$$\text{Area} \;=\; \int_0^H \Big[ x \Big]_{\frac{Ly}{H}}^L \mathrm{d}y \;=\; \int_0^H L\left(1 - \frac{y}{H}\right)\mathrm{d}y \;=\; L \Big[ y - \frac{y^2}{2H} \Big]_0^H \;=\; \tfrac{1}{2} HL$$

This illustrates that (for a well-behaved function) the order of integration in a multiple integral does not matter, and can be interchanged. The thing we do need to be careful about is the limits, because these usually need to be altered; for example, $x$ varied between 0 and $L$ in eqn (12.4) but goes from $Ly/H$ to $L$ in eqn (12.5). The best way of ascertaining the correct values for the limits is to sketch a diagram of the region being considered.

Although the order in which a multiple integral is carried out is up to us, the alternatives may entail different amounts of algebraic difficulty. As a rule-of-thumb, it's normally advisable to do the easiest integral first and leave the hardest until the end.

## 12.3 Choice of coordinates

As a slightly more interesting variant on the triangle of the previous section, let's derive the formula for the area of a circle. By symmetry, we can reduce the problem to being one of four times the area of $x^2 + y^2 \leqslant R^2$ in the positive quadrant. Integrating first along $y$, and then adding together all the narrow vertical strips in the $x$-direction, we have

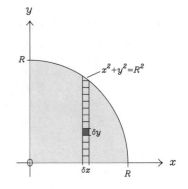

$$\text{Area of circle} = 4 \int_{x=0}^{x=R} dx \int_{y=0}^{y=\sqrt{R^2-x^2}} dy = 4 \int_{0}^{R} \sqrt{R^2 - x^2} \, dx \quad (12.6)$$

While the $y$-integral is straightforward, the second step is less so (requiring the substitution $x = R \sin \theta$). Reversing the order of integration does not help, but changing from Cartesian to polar coordinates does; rather than working in $x$ and $y$, we can tackle the problem in $r$ and $\theta$ (as in section 8.7).

Remembering that we are trying to sum up small elements of area, the box generated when the polar coordinates are varied by a tiny amount has sides of length $\delta r$ and $r \, \delta \theta$ (where the angle is in radians). In other words, we have

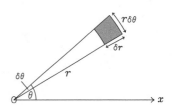

$$d\text{Area} = dx \, dy = r \, dr \, d\theta \quad (12.7)$$

With this relationship, we can then write eqn (12.6) as

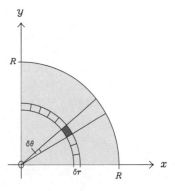

$$\text{Area of circle} = 4 \int_{r=0}^{r=R} r \, dr \int_{\theta=0}^{\theta=\frac{\pi}{2}} d\theta = 4 \left[ \tfrac{1}{2} r^2 \right]_0^R \left[ \theta \right]_0^{\frac{\pi}{2}} \quad (12.8)$$

where the range $\theta = 0$ to $\pi/2$ and $r = 0$ to $R$ covers the positive quadrant, irrespective of the order of integration. In eqn (12.8), the elementary boxes are first stacked up around the circle at a fixed radius; then these thin rings are added together, going from the centre to the circumference. The two integrals in eqn (12.8) are easily evaluated, and lead to the familiar result that the area of the circle is $\pi R^2$.

The example above highlights the point that multiple integrals, like many other mathematical operations, become much simpler if they are formulated in a coordinate system which matches the geometry of the problem. Although all our il-lustrations have been of double integrals, the generalization to triple, and higher-order, ones is straightforward; in particular, the conclusions regarding the order of the integration and the choice of coordinates holds equally well. Thus Carte-sian, cylindrical polar and spherical polar coordinates are best for problems hav-ing rectangular, cylindrical, and spherical characteristics respectively. Apart from the limits of the integrals, the only other thing we need to be careful about is the correct transformation of the elements of area, volume, and so on. This is done most easily by drawing a diagram, as we did for two-dimensional polars, but can also be calculated by a *Jacobian* determinant:

$$\mathrm{d}^M\mathrm{Vol} = \mathrm{d}x_1\,\mathrm{d}x_2\ldots\mathrm{d}x_M = |J|\,\mathrm{d}X_1\,\mathrm{d}X_2\ldots\mathrm{d}X_M \tag{12.9}$$

$$\mathrm{d}^3\mathrm{Vol} = \mathrm{d}x\,\mathrm{d}y\,\mathrm{d}z \quad \text{(Cartesian)}$$
$$= r\,\mathrm{d}r\,\mathrm{d}\theta\,\mathrm{d}z \quad \text{(Cylindrical)}$$
$$= r^2\sin\theta\,\mathrm{d}r\,\mathrm{d}\theta\,\mathrm{d}\phi \quad \text{(Spherical)}$$

$$\text{where} \quad J = \det\begin{pmatrix} \frac{\partial x_1}{\partial X_1} & \frac{\partial x_1}{\partial X_2} & \cdots & \frac{\partial x_1}{\partial X_M} \\ \frac{\partial x_2}{\partial X_1} & \frac{\partial x_2}{\partial X_2} & \cdots & \frac{\partial x_2}{\partial X_M} \\ \vdots & \vdots & \ddots & \vdots \\ \frac{\partial x_M}{\partial X_1} & \frac{\partial x_M}{\partial X_2} & \cdots & \frac{\partial x_M}{\partial X_M} \end{pmatrix} \tag{12.10}$$

where M is the dimension of the 'volume' (so that $M = 2$ is an area), $\{x_i\}$ are Cartesian components ($x_1 = x$, $x_2 = y$, $x_3 = z$, etc.) and $\{X_j\}$ are the alternative coordinates (e.g. $X_1 = r$, $X_2 = \theta$, and so on). In the two-dimensional polar trans-formation, for example, $x = r\cos\theta$ and $y = r\sin\theta$; eqn (12.10) then returns a Jacobian of $r$, consistent with eqn (12.7).

With the material in this section, we are able to do the integral of a *Gaussian* function which was beyond us in chapter 5

$$I = \int_0^\infty \mathrm{e}^{-x^2}\,\mathrm{d}x \tag{12.11}$$

The first step is non-intuitive, and consists of squaring $I$

$$I^2 = \int_0^\infty\int_0^\infty \mathrm{e}^{-(x^2+y^2)}\,\mathrm{d}x\,\mathrm{d}y \tag{12.12}$$

where we have multiplied eqn (12.11) with a copy of itself, but using $y$ as the dummy variable (instead of $x$). Changing from Cartesian to polar coordinates, according to eqn (12.7), eqn (12.12) becomes

$$\int_{-\infty}^\infty \mathrm{e}^{-\frac{1}{2}x^2/\sigma^2}\,\mathrm{d}x = \sigma\sqrt{2\pi}$$

$$I^2 = \int_{r=0}^\infty r\,\mathrm{e}^{-r^2}\,\mathrm{d}r \int_{\theta=0}^{\pi/2}\mathrm{d}\theta = \left[-\tfrac{1}{2}\mathrm{e}^{-r^2}\right]_0^\infty \left[\theta\right]_0^{\frac{\pi}{2}}$$

where the new limits still pertain to the positive quadrant. The product of the (now) two easy integrals yields $I^2 = \pi/4$; hence, on taking square roots, we find that eqn (12.11) is equal to $\sqrt{\pi}/2$.

## 12.4 **Worked examples**

**(1)** Show that the area of the ellipse $\dfrac{x^2}{a^2} + \dfrac{y^2}{b^2} = 1$ is equal to $\pi a b$.

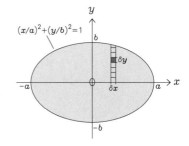

$$\text{Area of ellipse} = \underbrace{\iint_{\text{ellipse}} \underbrace{\text{dArea}}_{\text{d}x\,\text{d}y}}_{} = 4 \int_{x=0}^{a} \text{d}x \int_{y=0}^{b\sqrt{1-(x/a)^2}} \text{d}y = 4b \int_0^a \sqrt{1 - \frac{x^2}{a^2}}\ \text{d}x$$

Putting $x = a \sin \theta$,

$$\text{Area} = 4b \int_{\theta=0}^{\pi/2} \cos \theta \underbrace{a \cos \theta\ \text{d}\theta}_{\text{d}x} = 4ab \int_0^{\pi/2} \tfrac{1}{2}(\cos 2\theta + 1)\ \text{d}\theta$$

$$\therefore\ \text{Area of ellipse} = 2ab\left[\tfrac{1}{2}\sin 2\theta + \theta\right]_0^{\frac{\pi}{2}} = \underline{\underline{\pi a b}}$$

Early in the calculation, we replaced the integral over the ellipse by four times that of the positive quadrant; such manipulations are often useful in problems with a lot of symmetry, as they help to avoid subsequent difficulties arising from the limits of the integrals. It is also good practice to check derivations by making sure that they lead to sensible, or well-known, results under simplifying conditions. For example, an ellipse becomes a circle when $a = b$; our formula then correctly reduces to the familiar form $\pi a^2$, where $a$ is the radius.

Incidentally, the perimeter of an ellipse, which can easily be written down as a line integral ($\int \text{d}L$) using the parametric form $x = a \cos \theta$ and $y = b \cos \theta$ with $0 \leqslant \theta \leqslant 2\pi$, does not reduce to a nice simple formula; in fact, the results have to be calculated numerically or looked up in a table of *elliptical integrals*.

$$L = \int_0^{2\pi} \sqrt{a^2 \sin^2\theta + b^2 \cos^2\theta}\ \text{d}\theta$$

**(2)** Drawing a suitable diagram, show that an element of volume in spherical polar coordinates is given by $r^2 \sin \theta\ \text{d}r\ \text{d}\theta\ \text{d}\phi$; hence derive the formula for the volume of a sphere of radius $R$.

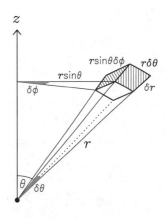

$$\text{Volume of small cuboid,}\quad \delta\text{Vol} \approx \delta r \times r\,\delta\theta \times r\sin\theta\,\delta\phi$$

$$\therefore\ \text{Volume of infinitesimal element,}\quad \text{dVol} = r^2 \sin\theta\ \text{d}r\ \text{d}\theta\ \text{d}\phi$$

$$\text{Volume of sphere} = \iiint_{\text{sphere}} \text{dVol} = \int_{r=0}^{R} r^2\,\text{d}r \int_{\theta=0}^{\pi} \sin\theta\,\text{d}\theta \int_{\phi=0}^{2\pi} \text{d}\phi$$

$$= \left[\tfrac{1}{3}r^3\right]_0^R \left[-\cos\theta\right]_0^{\pi} \left[\phi\right]_0^{2\pi} = \underline{\underline{\tfrac{4}{3}\pi R^3}}$$

Although this triple integral is easy to evaluate, because spherical polar coordinates match the natural geometry of the object and the problem reduces to a simple product of three one-dimensional integrals, we will see an alternative way of doing the calculation in the next example which exploits the symmetry of the sphere from the outset.

Incidentally, our diagram also enables us to write down the expression for an element of surface area for a sphere of radius $R$ and, thereby, derive the formula for the surface area of a sphere:

$$\delta S \approx R\,\delta\theta \times R\sin\theta\,\delta\phi$$

$$\text{Surface area} = \underbrace{\iint_{\text{sphere}} R^2 \sin\theta\,\mathrm{d}\theta\,\mathrm{d}\phi}_{\mathrm{d}S} = R^2 \int_{\theta=0}^{\pi} \sin\theta\,\mathrm{d}\theta \int_{\phi=0}^{2\pi} \mathrm{d}\phi = \underline{4\pi R^2}$$

> **(3)** Using cylindrical polar coordinates, show that the volume of the solid generated when the curve $y = f(x)$, for $a \leqslant x \leqslant b$, is rotated about the $x$-axis through $360°$ is given by $\pi\int_a^b y^2\,\mathrm{d}x$.

$$\text{Volume of infinitesimal element,}\quad \mathrm{d}\text{Vol} = r\,\mathrm{d}r\,\mathrm{d}\theta\,\mathrm{d}x$$

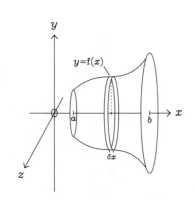

$$\therefore\quad \text{Volume of revolution} = \iiint_{\text{solid}} \mathrm{d}\text{Vol} = \int_{x=a}^{b} \mathrm{d}x \int_{r=0}^{f(x)} r\,\mathrm{d}r \int_{\theta=0}^{2\pi} \mathrm{d}\theta$$

$$= 2\pi \int_{x=a}^{b} \left[\tfrac{1}{2}r^2\right]_{r=0}^{f(x)} \mathrm{d}x$$

$$\text{i.e.}\quad \text{Volume of revolution} = \pi \int_a^b \left[f(x)\right]^2\,\mathrm{d}x = \pi \int_a^b \underline{y^2\,\mathrm{d}x}$$

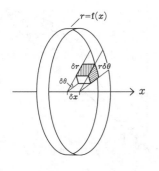

Notice that, unlike in the previous example, the triple integral doesn't reduce to the product of three one-dimensional integrals. This is because the radius, or the maximum value of the $r$ component, depends on the $x$-coordinate: $r = f(x)$. Thus, while the angular $\theta$-integral is self-contained, the radial and lateral contributions are interlinked through the limits; indeed, the sum is very difficult to formulate if the order of the $r$ and $x$ integrals is reversed.

The initial evaluation of the angular and radial integrals has a simple geometrical interpretation in that it represents the volume of a thin circular disc at a given $x$-coordinate: $\pi\left[f(x)\right]^2\,\mathrm{d}x$. Hence, the expression for the volume of revolution represents the sum of many such thin discs stacked up from $x = a$ to $x = b$.

Finally, since a sphere of radius $R$ can be generated by rotating the semi-circular arc $x^2 + y^2 = R^2$ (for $y \geqslant 0$) through $360°$, the formula of the previous example could also be obtained as follows:

$$\text{Volume of sphere} \;=\; \pi \int_{-R}^{R} (R^2 - x^2)\, dx \;=\; \pi \left[ R^2 x - \tfrac{1}{3} x^3 \right]_{-R}^{R} \;=\; \tfrac{4}{3} \pi R^3$$

**(4)** In spherical polar coordinates, the $2p$ orbitals of hydrogen are of the form

$$\psi_j(r, \theta, \phi) \;=\; A_j \, r\, e^{-\frac{1}{2} r / a_o} \, f_j(\theta, \phi)$$

where $a_o$ is the Bohr radius and the index $j$ denotes $x$ or $y$ or $z$. The angular functions are $f_x = \sin\theta \cos\phi$, $f_y = \sin\theta \sin\phi$, and $f_z = \cos\theta$. Find the normalization constants, $A_j$, to ensure $\iiint |\psi_j|^2 \, d\text{Vol} = 1$ and show that $\iiint \psi_j \psi_k \, d\text{Vol} = 0$ for $j \neq k$.

$$\underset{\text{All space}}{\iiint} |\psi_j|^2 \, d\text{Vol} \;=\; \underset{\text{All space}}{\iiint} A_j^2 \, r^2 \, e^{-r/a_o} \left[ f_j(\theta, \phi) \right]^2 \underbrace{r^2 \sin\theta \, dr \, d\theta \, d\phi}_{d\text{Vol}}$$

$$=\; A_j^2 \int_{r=0}^{\infty} r^4 \, e^{-r/a_o} \, dr \int_{\theta=0}^{\pi} \sin\theta \, d\theta \int_{\phi=0}^{2\pi} \left[ f_j(\theta, \phi) \right]^2 \, d\phi \;=\; 1$$

But $\quad \displaystyle\int_{r=0}^{\infty} r^4 \, e^{-r/a_o} \, dr \;=\; \underbrace{\int_{t=0}^{\infty} (a_o t)^4 \, e^{-t} \, a_o \, dt}_{r = a_o t} \;=\; a_o^5 \underbrace{\int_{0}^{\infty} t^4 \, e^{-t} \, dt}_{4!}$  $\qquad$ [See section 5.7]

$$\therefore \quad 24 \, a_o^5 \, A_j^2 \underbrace{\int_{\theta=0}^{\pi} \sin\theta \, d\theta \int_{\phi=0}^{2\pi} \left[ f_j(\theta, \phi) \right]^2 \, d\phi}_{I_j} \;=\; 1 \qquad -(1)$$

Where $\quad \displaystyle I_x \;=\; \int_{\theta=0}^{\pi} \sin^3\theta \, d\theta \int_{\phi=0}^{2\pi} \cos^2\phi \, d\phi$

$$=\; \int_{0}^{\pi} \underbrace{(1 - \cos^2\theta)}_{\sin^2\theta} \sin\theta \, d\theta \int_{0}^{2\pi} \tfrac{1}{2} (\cos 2\phi + 1) \, d\phi$$

$$=\; \underbrace{\left[ \tfrac{1}{3} \cos^3\theta - \cos\theta \right]_{0}^{\pi}}_{4/3} \underbrace{\left[ \tfrac{1}{4} \sin 2\phi + \tfrac{1}{2} \phi \right]_{0}^{2\pi}}_{\pi} \;=\; \frac{4\pi}{3}$$

Similarly $\quad \displaystyle I_y \;=\; \int_{\theta=0}^{\pi} \sin^3\theta \, d\theta \int_{\phi=0}^{2\pi} \sin^2\phi \, d\phi \;=\; \frac{4\pi}{3}$

And $\quad I_z = \int\limits_{\theta=0}^{\pi} \cos^2\theta \sin\theta \, d\theta \int\limits_{\phi=0}^{2\pi} d\phi = \underbrace{\left[-\tfrac{1}{3}\cos^3\theta\right]_0^{\pi}}_{2/3} \underbrace{\left[\phi\right]_0^{2\pi}}_{2\pi} = \dfrac{4\pi}{3}$

Since $I_x = I_y = I_z = \dfrac{4\pi}{3}$, $\quad$ (1) $\Rightarrow$ $\quad A_x = A_y = A_z = \dfrac{1}{4a_o^2\sqrt{2\pi a_o}}$

$\therefore \quad \iiint\limits_{\text{All space}} \psi_j\,\psi_k \, d\text{Vol} = \dfrac{I_{jk}}{32\pi a_o^5} \int\limits_0^{\infty} r^4\, e^{-r/a_o}\, dr = \dfrac{3I_{jk}}{4\pi} \quad\quad - (2)$

where $\quad I_{jk} = \int\limits_{\theta=0}^{\pi} \sin\theta \, d\theta \int\limits_{\phi=0}^{2\pi} f_j(\theta,\phi)\, f_k(\theta,\phi)\, d\phi$

But $\quad I_{xy} = \int\limits_{\theta=0}^{\pi} \sin^3\theta \, d\theta \int\limits_{\phi=0}^{2\pi} \cos\phi \sin\phi \, d\phi \propto \underbrace{\left[\tfrac{1}{2}\sin^2\phi\right]_0^{2\pi}}_{0} = 0$

$I_{xz} = \int\limits_{\theta=0}^{\pi} \sin^2\theta \cos\theta \, d\theta \int\limits_{\phi=0}^{2\pi} \cos\phi \, d\phi \propto \underbrace{\left[\sin\phi\right]_0^{2\pi}}_{0} = 0$

$I_{yz} = \int\limits_{\theta=0}^{\pi} \sin^2\theta \cos\theta \, d\theta \int\limits_{\phi=0}^{2\pi} \sin\phi \, d\phi \propto \underbrace{\left[-\cos\phi\right]_0^{2\pi}}_{0} = 0$

Hence $\quad$ (2) $\Rightarrow$ $\quad \iiint\limits_{\text{All space}} \psi_j\,\psi_k \, d\text{Vol} = 0 \quad$ for $\ j \neq k$

When a multiple integral factorizes into a product of terms, it's worth trying to spot whether any of them are zero before undertaking the whole sum. If one of the angular factors is zero, that integral should be done first because the evaluation of the others then becomes redundant. In the second part of the example above, the result hinges on the integration over the angle $\phi$.

# 13 Ordinary differential equations

## 13.1 Definition of terms

An *ordinary* differential equation (ODE) is one that involves only normal differentials of $y(x)$, that is $dy/dx$ and its derivatives, and none of those partial curly-d's which we discussed in chapter 10. The *order* of a differential equation is the highest derivative of $y$ that appears in it: a first-order differential equation includes terms like $y' = \frac{dy}{dx}$ but no higher-order ones, such as $y'' = \frac{d^2 y}{dx^2}$ or $y''' = \frac{d^3 y}{dx^3}$. We have already met the concept of *linearity* in chapter 4 which, when applied to differential equations, means that $y$ and its derivatives are raised to no power higher than the first; equations where this is not true are called *non-linear*. The *degree* of a differential equation is the power to which the highest derivative (of $y$) is raised, but we will only be dealing with those of the first.

## 13.2 First-order: separable

The most elementary ODE is

$$\frac{dy}{dx} = k \tag{13.1}$$

with $k$ a constant, and is first-order, linear, and of first-degree. As we learnt in chapter 4, this equation describes a function $y(x)$ which has a constant slope; so, its solution is simply a straight line. We can find the equation of this line by integrating both sides of eqn (13.1) with respect to $x$, giving

$$y = kx + A \tag{13.2}$$

Note that a single arbitrary constant $A$ has appeared as a result of a single integration, and this introduces a result of crucial importance – the *general solution* of an $n^{\text{th}}$-order ODE must contain $n$ arbitrary constants. It is easy to see why one, and only one, arbitrary constant is needed in this first-order case: eqn (13.1) tells us that $y(x)$ has a constant slope $k$, but contains nothing about the intercept of the associated line with $x = 0$. Hence any $A$ will satisfy eqn (13.1), and we need a *boundary condition* to fix its value. For example, if we were told that $y = 0$ when $x = 0$ then we would know that $A = 0$, and would have the unique solution of a straight line of slope $k$ passing through the origin.

A more general first-order ODE is

$$\frac{dy}{dx} = X(x)\,Y(y) \tag{13.3}$$

in which $X$ and $Y$ are general functions of $x$ and $y$ respectively. Just like eqn (13.1), this is a *separable* equation because, dividing both sides by $Y$ and integrating with respect to $x$, we obtain

$$\int \frac{dy}{Y(y)} = \int X(x)\,dx \tag{13.4}$$

which can be used to get a solution for $y(x)$. Let's illustrate how this works: if $X(x) = \cos x$ and $Y(y) = y$, then eqn (13.4) gives

$$\frac{dy}{dx} = y\cos x$$

$$\ln y = \sin x + A, \quad \text{or} \quad y = Be^{\sin x}$$

where an additive constant has been replaced by a multiplicative one, $B = e^{A}$, on taking exponentials.

## 13.3 First-order: homogeneous

A function $f(x, y)$ is *homogeneous*, and of degree $m$, if it scales by a factor $\lambda^{m}$ when both $x$ and $y$ are multiplied by $\lambda$; $f = x^{3} + 3x^{2}y + 3y^{2}x + 5y^{3}$, for example, which is a third-degree homogeneous function. It can be rewritten as a product of $x^{m}$ and a function of the ratio $\frac{y}{x}$: $f = x^{m}\,F(v)$ where $y = vx$. In our example, $F(v) = 1 + 3v + 3v^{2} + 5v^{3}$. Now, consider the ODE

$$\frac{dy}{dx} = \frac{\theta(x,y)}{\phi(x,y)} \tag{13.5}$$

where $\theta(x,y)$ and $\phi(x,y)$ are both homogeneous and of the same degree. The derivative can then be expressed as a function of $v$ only:

$$\frac{dy}{dx} = \frac{x^{m}\,\Theta(v)}{x^{m}\,\Phi(v)} = \psi(v) \tag{13.6}$$

Differentiating $y = vx$ with respect to $x$, using the product rule of eqn (4.9), we obtain a separable ODE in terms of $v$ and $x$

$$\frac{dy}{dx} = v + x\frac{dv}{dx} = \psi(v) \tag{13.7}$$

which can be re-arranged into the form of eqn (13.4)

$$\int \frac{dv}{\psi(v) - v} = \int \frac{dx}{x}$$

and hence solved for $v$. The final answer is ascertained by using $y = vx$, and a suitable boundary condition.

There are various 'change-of-variable tricks' which can transform some *inhomogeneous* first-order ODEs into a soluble homogeneous form. Consider, for example, the inhomogeneous case

$$\frac{dy}{dx} = \frac{x + y + 1}{x - y}$$

By putting $x = X + a$ and $y = Y + b$, and choosing the constants $a$ and $b$ so that $a + b + 1 = 0$ and $a - b = 0$, the ODE above becomes

$$\frac{dy}{dx} = \frac{dY}{dX} = \frac{X + Y}{X - Y}$$

$$\delta x = \delta X \quad \text{and} \quad \delta y = \delta Y$$

which is homogeneous in terms of $X$ and $Y$. This can be solved with the substitution $Y = vX$, as before, with $a = b = \frac{1}{2}$.

$$v + X\frac{dv}{dX} = \frac{1 + v}{1 - v}$$

## 13.4 First-order: integrating factor

A general linear first-order ODE is of the type

$$\frac{dy}{dx} + y\,P(x) = Q(x) \tag{13.8}$$

and can be solved by multiplying both sides by an *integrating factor*, $\mathcal{I}(x)$,

$$\mathcal{I}(x) = \exp\left[\int P(x)\,dx\right] \tag{13.9}$$

Let's take as an example the equation

$$\frac{dy}{dx} + \frac{y}{x} = x$$

$$P(x) = \frac{1}{x}, \quad Q(x) = x$$

In this case $\mathcal{I}(x) = \exp\left[\int dx/x\right] = \exp[\ln x] = x$, so that we get

$$x\frac{dy}{dx} + y = \frac{d}{dx}[xy] = x^2 \quad \Rightarrow \quad y = x^{-1}\left(\tfrac{1}{3}x^3 + A\right)$$

The critical step here is that '$\mathcal{I}(x)$ times the left-hand side of the ODE can be rewritten as the derivative of $\mathcal{I}(x)$ times $y$'. It is also important to note that the arbitrary constant $A$ is not simply tacked onto the end of the answer for $y$.

$$\frac{d}{dx}\Big[\mathcal{I}(x)\,y(x)\Big] = \mathcal{I}(x)\,Q(x)$$

The formula of eqn (13.9) can be derived by equating the derivative of the product $\mathcal{I}(x)\,y(x)$ to $\mathcal{I}(x)$ times the left-hand side of eqn (13.8):

$$\frac{d}{dx}[\mathcal{I}y] = \mathcal{I}\frac{dy}{dx} + y\frac{d\mathcal{I}}{dx} = \mathcal{I}\frac{dy}{dx} + \mathcal{I}y\,P(x)$$

$$\frac{d\mathcal{I}}{dx} = \mathcal{I}\,P(x)$$

This simplifies to give a separable first-order ODE

$$\int\frac{d\mathcal{I}}{\mathcal{I}} = \int P(x)\,dx \tag{13.10}$$

which can easily be solved to yield eqn (13.9).

## 13.5 Second-order: homogeneous

Now let's consider some second-order linear ODEs. We will largely restrict our discussion to the situation where the coefficients are constants, rather than being functions of $x$. In the simplest case, the right-hand side is zero and the ODE is said to be homogeneous:

$$\frac{d^2y}{dx^2} + k_1\frac{dy}{dx} + k_2 y = 0 \tag{13.11}$$

where $k_1$ and $k_2$ are constants. We can guess that the solutions of this equation are of the form $y = Ae^{\alpha x}$, where $A$ and $\alpha$ are constants, because this is the only function that retains the same form on repeated differentiation. Substitution of this trial solution into eqn (13.11) leads to the requirement that

$$Ae^{\alpha x}\left(\alpha^2 + k_1\alpha + k_2\right) = 0 \tag{13.12}$$

Apart from the trivial $y=0$ case, when $A=0$, the permissible values of $\alpha$ for the exponential solutions are given by the roots of the quadratic 'auxiliary equation'. There are three cases of interest.

Case (i) is when there are two different real roots for $\alpha$. For example,

Try $y = Ae^{\alpha x}$

$$\frac{d^2y}{dx^2} - \omega_0^2 y = 0 \quad\Rightarrow\quad \alpha^2 - \omega_0^2 = 0, \quad\therefore\quad \alpha = \pm\omega_0 \tag{13.13}$$

This means there are two distinct solutions

$$y_1 = A_1 e^{+\omega_0 x} \quad\text{and}\quad y_2 = A_2 e^{-\omega_0 x}$$

where the constants $A_1$ and $A_2$ have to be determined from two boundary conditions. Before imposing the latter, however, we should write the general solution as the sum of $y_1$ and $y_2$ due to the linearity of the ODE:

$$y = Ae^{+\omega_0 x} + Be^{-\omega_0 x} \tag{13.14}$$

where we have relabelled the constants as $A$ and $B$ to avoid subscripts. Note that it is equally valid to express the general solution of eqn (13.13) as

$C = A - B$

$D = A + B$

$$y = C\sinh(\omega_0 x) + D\cosh(\omega_0 x) \tag{13.15}$$

because of the hyperbolic identities of eqn (7.18).

Case (ii) is when there are imaginary roots for $\alpha$. For example,

$$\frac{d^2y}{dx^2} + \omega_0^2 y = 0 \quad\Rightarrow\quad \alpha^2 + \omega_0^2 = 0, \quad\therefore\quad \alpha = \pm i\omega_0 \tag{13.16}$$

where $i^2 = -1$. This is the equation of motion for a *simple harmonic* oscillator, with $y$ being displacement and $x$ time, and is used as an analytical model for many physical problems. The exponential solution

$$y = Ae^{+i\omega_0 x} + Be^{-i\omega_0 x}$$

can also be expressed in terms of trigonometric functions

$$y = C\sin(\omega_0 x) + D\cos(\omega_0 x) = F\sin(\omega_0 x + \phi) \quad (13.17)$$

because of eqn (7.11) and the material in chapter 3. Which form is best depends on the nature of the boundary conditions but, as the associated constants are all related, there is no fundamental difference.

A non-zero coefficient for $dy/dx$ gives a conjugate pair of complex roots for $\alpha$, and a solution that has characteristics of both eqns (13.14) and (13.17)

$$\frac{d^2y}{dx^2} + \frac{dy}{dx} + y = 0 \quad \Rightarrow \quad y = F\sin\left(\frac{\sqrt{3}}{2}x + \phi\right)e^{-x/2} \quad (13.18)$$

This example shows why the associated harmonic system is described as being *damped*: the oscillation is multiplied by a decaying exponential term which goes to zero as time $(x)$ tends to infinity.

Finally, case (iii) is where the roots of the quadratic are equal; for example

$$\frac{d^2y}{dx^2} + 2\frac{dy}{dx} + y = 0 \quad \Rightarrow \quad y = Ae^{-x} + Be^{-x} = Ce^{-x} \quad (13.19)$$

This method of solution produces only one arbitrary constant $(C = A + B)$ and, so, a second functional form must be guessed. The correct form is $y = xe^{-\alpha x}$, which can be seen to give zero on substitution into the ODE of eqn (13.19). Hence, the general solution is

$$y = (C + Dx)e^{-x}$$

In the context of the ubiquitous harmonic oscillator, this situation is often called *critical damping*.

Some second-order linear ODEs without constant coefficients can be transformed into the simple case of eqn (13.11) via a suitable change of variable. Consider, for example

$$x^2\frac{d^2y}{dx^2} + x\frac{dy}{dx} + y = 0 \quad (13.20)$$

Putting $u = \ln(x)$, and using the chain rule of eqn (4.12), the transformation of the first derivative is straightforward

$$\frac{dy}{dx} = \frac{dy}{du}\frac{du}{dx} = \frac{1}{x}\frac{dy}{du} \quad (13.21)$$

For the second derivative

$$\frac{d^2y}{dx^2} = \frac{d}{dx}\left(\frac{1}{x}\frac{dy}{du}\right) = \frac{1}{x}\frac{d}{dx}\left(\frac{dy}{du}\right) - \frac{1}{x^2}\frac{dy}{du} = \frac{1}{x^2}\left[\frac{d}{du}\left(\frac{dy}{du}\right) - \frac{dy}{du}\right]$$

which leads to

$$x^2\frac{d^2y}{dx^2} = \frac{d^2y}{du^2} - \frac{dy}{du} \quad (13.22)$$

Substitution from eqns (13.21) and (13.22) into eqn (13.20) transforms it into a second-order ODE with constant coefficients in terms of $y$ and $u$.

$$F\cos\phi = C = i(A - B)$$
$$F\sin\phi = D = A + B$$

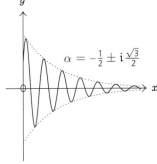

$$\alpha = -\tfrac{1}{2} \pm i\frac{\sqrt{3}}{2}$$

$$\alpha = -1 \text{ (twice)}$$

$$\left[\alpha^2 + 2\alpha + 1 = (\alpha + 1)^2 = 0\right]$$

$$\frac{d}{dx} = \frac{1}{x}\frac{d}{du}$$

$$\frac{d^2y}{du^2} + y = 0$$

## 13.6 Second-order: inhomogeneous

Now let's consider the inhomogeneous case, when the right-hand side of eqn (13.11) is non-zero:

$$\frac{d^2y}{dx^2} + k_1 \frac{dy}{dx} + k_2 y = F(x) \tag{13.23}$$

where $F(x)$ is any reasonable function of $x$. The linearity of the ODE means that the general solution can be expressed as the sum of a *complementary function* $c(x)$ and a *particular integral* $p(x)$:

$$y(x) = c(x) + p(x)$$

where $c(x)$ is the general solution of eqn (13.11) and $p(x)$ satisfies eqn (13.23). Since the homogeneous solution was discussed in the preceding section, we'll concentrate here on finding $p(x)$; it should have no arbitrary constants as $c(x)$ has the requisite number.

There is an easy recipe for finding a suitable $p(x)$ for a given $F(x)$, although the algebra can be messy: try $p(x)$ being a linear superposition of all the types of functions produced by $F(x)$ and its first and second derivatives. The values of the related coefficients are then determined by equating the pre-factors of each functional type on each side of eqn (13.23).

An example should help to make this clear:

$$\frac{d^2y}{dx^2} + \omega_0^2 y = x \sin x \tag{13.24}$$

for $\omega_0 \neq 1$. Following eqn (13.16), $c(x) = A\sin(\omega_0 x) + B\cos(\omega_0 x)$ and we need only to find $p(x)$. Our recipe suggests we try

$$p(x) = Dx\sin x + Ex\cos x + G\sin x + H\cos x$$

where $D$, $E$, $G$, and $H$ are the constants that define the linear combination of the functions that arise from the zeroth, first, and second derivatives of $x\sin x$. Substituting into eqn (13.24), and equating coefficients of the various functional types, produces the simultaneous equations

$$
\begin{aligned}
x\sin x: &\qquad D(\omega_0^2 - 1) = 1 \\
x\cos x: &\qquad E(\omega_0^2 - 1) = 0 \\
\sin x: &\qquad -2E + G(\omega_0^2 - 1) = 0 \\
\cos x: &\qquad 2D + H(\omega_0^2 - 1) = 0
\end{aligned}
$$

which give $D = (\omega_0^2 - 1)^{-1}$, $E = G = 0$ and $H = -2(\omega_0^2 - 1)^{-2}$. Hence, the general solution of eqn (13.24) is

$$y = A\sin(\omega_0 x) + B\cos(\omega_0 x) + \frac{x\sin x}{\omega_0^2 - 1} - \frac{2\cos x}{(\omega_0^2 - 1)^2}$$

where two boundary conditions are now needed to fix $A$ and $B$.

Physically, eqn (13.23) represents an oscillating system that is driven by a force $F(x)$. A damping term is present in real situations, so that the complementary function decays to zero as $x$ (time) increases. This part of the solution, which is the only one that depends on the boundary conditions, is said to be *transient*; the particular integral gives the *steady-state* response.

Let us conclude this chapter by mentioning a difficulty sometimes encountered with eqn (13.23). Consider the seemingly innocuous case $k_1 = 0$, $k_2 = 1$, and $F(x) = \sin x$. The complementary function is the solution of eqn (13.16) with $\omega_0 = 1$; namely, $c(x) = A \sin x + B \cos x$. Unfortunately our recipe for the particular integral gives $p(x) = D \sin x + E \cos x$ which is the same as the complementary function, and produces zero when substituted into eqn (13.23). How then do we produce non-vanishing $\sin x$ terms to match those in $F(x)$? Well, we need to guess functions that produce terms in $\sin x$ on differentiation. A reasonable first guess is $p(x) = D x \sin x + E x \cos x$. This trial solution yields $D = 0$ and $E = -1/2$, so that $y = A \sin x + B \cos x - \frac{1}{2} x \cos x$.

$$\frac{d^2 y}{dx^2} + y = \sin x$$

## 13.7 Worked examples

---

**(1)** The number of radioactive atoms in a sample, $N$, decays with time, $t$, according to the law $dN/dt = -\lambda N$. Solve the equation for $N(t)$ and obtain an expression for the *half-life*, $\tau$, the time taken for the radioactivity to fall by 50%.

---

$$\frac{dN}{dt} = -\lambda N \quad \Rightarrow \quad \int \frac{dN}{N} = -\lambda \int dt$$

$$\therefore \quad \ln N = A - \lambda t$$

If $N = N_0$ when $t = 0$, $A = \ln N_0$.

$$\therefore \quad \ln N - \ln N_0 = \ln \left[ \frac{N}{N_0} \right] = -\lambda t \quad \Rightarrow \quad \underline{N = N_0 e^{-\lambda t}}$$

Half-life $\tau \quad \Rightarrow \quad \dfrac{N(t+\tau)}{N(t)} = \dfrac{1}{2}$

$$\therefore \quad e^{-\lambda \tau} = \tfrac{1}{2} \quad \Leftrightarrow \quad -\lambda \tau = \ln\left(\tfrac{1}{2}\right) = -\ln 2$$

$$\therefore \quad \underline{\tau = \tfrac{\ln 2}{\lambda}}$$

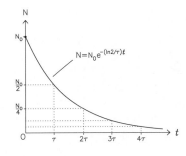

The radioactivity measured by a Geiger counter, $G(t)$ say, follows the $N(t)$ curve because they are proportional to each other. The simplest way to estimate the half-life of the sample is to plot the logarithm of the signal versus time, as this gives a straight line with a negative slope of $\lambda$. As shown above, the half-life is just $\tau \approx 0.693/\lambda$.

$$\ln G = \text{Constant} - \lambda t$$

**(2)** The rate equations for first- and second-order chemical reactions take the form

$$-\frac{dc}{dt} = kc \quad \text{and} \quad -\frac{dc}{dt} = kc^2$$

respectively, where $c$ is the concentration of the reactant, $t$ is time, and k is the rate constant. Solve these equations for $c(t)$ given $c(0) = c_0$, and explain how the two cases can be distinguished graphically given measurements of concentration versus time.

The first-order rate equation is the same as the radioactivity case in the previous example, with N replaced by $c$ and k instead of $\lambda$. Hence

$$c = c_0 e^{-kt} \quad \Longleftrightarrow \quad \ln c = \ln c_0 - kt \qquad -(1)$$

The second-order rate equation is also of the separable type discussed in section 13.2, so that

$$\int \frac{dc}{c^2} = -k \int dt \quad \Rightarrow \quad -\frac{1}{c} = A - kt$$

Applying the boundary condition, $c = c_0$ when $t = 0$, tells us that $A = -1/c_0$. This gives the solution

$$\frac{1}{c} = \frac{1}{c_0} + kt \qquad -(2)$$

Since linear trends are the easiest to pick out by eye, (1) and (2) indicate that we should plot graphs of the logarithm and the reciprocal of the concentration versus time. If $\ln c$ gives a straight line, then it's a first-order chemical reaction; if $1/c$ gives a straight line, then it's a second-order one.

---

**(3)** Obtain general solutions of the following first-order ODEs

(a) $\dfrac{dy}{dx} = \dfrac{1-y^2}{x}$    (b) $\dfrac{dy}{dx} = \dfrac{2y^2 + xy}{x^2}$    (c) $\dfrac{dy}{dx} = \dfrac{x+y+5}{x-y+2}$

(d) $\dfrac{dy}{dx} + y\cot x = \operatorname{cosec} x$    (e) $\dfrac{dy}{dx} + 2xy = x$    (f) $\dfrac{dy}{dx} + \dfrac{y}{x} = \cos x$

---

[See section 7.7]

(a) $\displaystyle\int \frac{dy}{1-y^2} = \int \frac{dx}{x} \quad \Rightarrow \quad \tanh^{-1} y = \ln x + A$

By using the exponential definition of the hyperbolic tangent, or by integrating with respect to $y$ using partial fractions, the solution $y = \tanh(\ln x + A)$ can also be expressed as

$B = e^{-2A}$

$$y = \frac{x^2 - B}{x^2 + B}$$

(b) $\dfrac{dy}{dx} = \dfrac{2y^2 + xy}{x^2} = 2v^2 + v \qquad$ where $\; y = vx$

$\therefore \; \dfrac{dy}{dx} = v + x\dfrac{dv}{dx} = 2v^2 + v$

$\therefore \; \displaystyle\int \dfrac{dv}{v^2} = 2\int \dfrac{dx}{x} \qquad \Rightarrow \qquad -\dfrac{1}{v} = 2\ln x + A$

$\therefore \; \underline{y = \dfrac{-x}{2\ln x + A}}$

(c) Putting $\; x = X - \tfrac{7}{2} \;$ and $\; y = Y - \tfrac{3}{2} \qquad\qquad\qquad\qquad a = -\tfrac{7}{2}, \; b = -\tfrac{3}{2}$

$\dfrac{dy}{dx} = \dfrac{x + y + 5}{x - y + 2} \qquad \Rightarrow \qquad \dfrac{dY}{dX} = \dfrac{X + Y}{X - Y} \qquad\qquad \begin{array}{l} a + b + 5 = 0 \\ a - b + 2 = 0 \end{array}$

Putting $\; Y = vX, \qquad v + X\dfrac{dv}{dX} = \dfrac{1 + v}{1 - v}$

$\therefore \; X\dfrac{dv}{dX} = \dfrac{1 + v}{1 - v} - v = \dfrac{1 + v^2}{1 - v}$

$\therefore \; \displaystyle\underbrace{\int \dfrac{dX}{X}}_{\ln X} = \int \dfrac{1 - v}{1 + v^2}\, dv = \underbrace{\int \dfrac{dv}{1 + v^2}}_{\tan^{-1} v} - \underbrace{\int \dfrac{v}{1 + v^2}\, dv}_{\frac{1}{2}\ln(1 + v^2)}$

$\therefore \; A + \ln X = \tan^{-1}\left(\tfrac{Y}{X}\right) - \tfrac{1}{2}\ln\left[1 + \left(\tfrac{Y}{X}\right)^2\right]$

But $\; \underbrace{\tfrac{1}{2}\ln(x^2)}_{\ln X} + \tfrac{1}{2}\ln\left[1 + \left(\tfrac{Y}{X}\right)^2\right] = \tfrac{1}{2}\ln\left[x^2 + Y^2\right] \qquad\qquad \ln\alpha + \ln\beta = \ln(\alpha\beta)$

$\therefore \; \underline{\tan^{-1}\left(\dfrac{y + 3/2}{x + 7/2}\right) = \tfrac{1}{2}\ln\left[\left(x + \tfrac{7}{2}\right)^2 + \left(y + \tfrac{3}{2}\right)^2\right] + A}$

(d) Integrating factor, $\; \mathcal{I}(x) = \exp\left[\displaystyle\int \cot x\, dx\right]$

But $\; \displaystyle\int \cot x\, dx = \int \dfrac{\cos x}{\sin x}\, dx = \ln[\sin x] + K$

$\therefore \; \mathcal{I}(x) \propto \sin x$

$\therefore \; \dfrac{d}{dx}\left(y\sin x\right) = \operatorname{cosec} x \, \sin x = 1 \qquad\qquad\qquad \operatorname{cosec} x = \dfrac{1}{\sin x}$

$\therefore \; y\sin x = \displaystyle\int dx = x + A \qquad \text{i.e.} \;\; \underline{y = \dfrac{x + A}{\sin x}}$

(e)   Integrating factor,   $\mathcal{I}(x) = \exp\left[\int 2x\,dx\right] \propto e^{x^2}$

$$\therefore \quad \frac{d}{dx}\left(y\,e^{x^2}\right) = x\,e^{x^2}$$

$$\therefore \quad y\,e^{x^2} = \int x\,e^{x^2}\,dx = \tfrac{1}{2}e^{x^2} + A \qquad \text{i.e. } \underline{y = \tfrac{1}{2} + A\,e^{-x^2}}$$

(f)   Integrating factor,   $\mathcal{I}(x) = \exp\left[\int x^{-1}\,dx\right] \propto e^{\ln x} = x$

$$\therefore \quad \frac{d}{dx}(x\,y) = x\cos x$$

$$\therefore \quad x\,y = \int x\cos x\,dx = x\sin x - \underbrace{\int \sin x\,dx}_{A\,-\,\cos x}$$

$$\therefore \quad \underline{y = \sin x + \frac{\cos x - A}{x}}$$

---

**(4)** Use the substitution $u = y^k$ to transform the *Bernoulli* equation

$$\frac{dy}{dx} + y\,P(x) = y^\alpha\,Q(x) \qquad (\dagger)$$

into a form soluble with an integrating factor, where k is a suitable constant. Solve it when $P(x) = Q(x) = x$ and $\alpha = 2$.

---

$$u = y^k \quad \Rightarrow \quad \frac{du}{dx} = k\,y^{k-1}\frac{dy}{dx}$$

$$\therefore \ (\dagger) \times k\,y^{k-1} \quad \Rightarrow \quad \frac{du}{dx} + u\,k\,P(x) = \underbrace{y^{k-1+\alpha}}_{=1?}k\,Q(x)$$

$$\therefore \ \underline{k = 1 - \alpha} \text{ for linear ODE}$$

For $P(x) = Q(x) = x$ and $\alpha = 2$, let $u = y^{-1}$

$$\therefore \quad \frac{du}{dx} - x\,u = -x \quad \Rightarrow \quad \mathcal{I}(x) = \exp\left[-\int x\,dx\right] \propto e^{-x^2/2}$$

$$\therefore \quad \frac{d}{dx}\left(u\,e^{-x^2/2}\right) = -x\,e^{-x^2/2}$$

$$\therefore \quad u\,e^{-x^2/2} = e^{-x^2/2} + A \quad \Rightarrow \quad \underline{u^{-1} = y = \frac{1}{1 + A\,e^{x^2/2}}}$$

As is often the case with non-linear ODEs, the character of the solution depends critically on the value of the integration constant $A$.

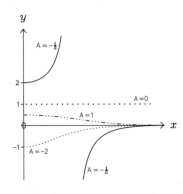

**(5)** Find the solutions of

(a) $y'' - 2y' - 3y = \sin x$, given $y(0) = 0$ and $y$ is finite as $x \to \infty$

(b) $y'' - 2y' - 8y = x^2 + e^{-2x}$     (c) $y'' + y' + y = \cos(\omega x)$

(a)  For complementary function, $c'' - 2c' - 3c = 0$, TRY $c = A e^{mx}$

$$\therefore \; m^2 - 2m - 3 = (m-3)(m+1) = 0$$

$$\therefore \; m = 3 \;\text{ or }\; m = -1 \;\;\Rightarrow\;\; c = A e^{-x} + B e^{3x}$$

For particular integral, TRY $p \;\;= \;\;\; D \sin x \;+\; E \cos x$

$$\therefore \; p' \;=\;\;\; D \cos x \;-\; E \sin x$$

$$\therefore \; p'' \;=\; -D \sin x \;-\; E \cos x$$

But $p'' - 2p' - 3p = \sin x$

$$\Rightarrow \;\; \underbrace{\left[-D + 2E - 3D\right]}_{1} \sin x + \underbrace{\left[-E - 2D - 3E\right]}_{0} \cos x = \sin x$$

$$\therefore \;\; \left. \begin{array}{l} -4D + 2E = 1 \\ \text{and} \;\; -2D - 4E = 0 \end{array} \right\} \;\; \begin{array}{l} D = -1/5 \\ E = 1/10 \end{array}$$

Therefore general solution is

$$y(x) = c(x) + p(x) = A e^{-x} + B e^{3x} - \tfrac{1}{5}\sin x + \tfrac{1}{10}\cos x$$

But $y$ finite as $x \to \infty \;\;\Rightarrow\;\; B = 0$

and $y(0) = 0 \;\;\Rightarrow\;\; A = -\tfrac{1}{10}$

Therefore solution is $\;\; y(x) = \tfrac{1}{10}\left(\cos x - 2\sin x - e^{-x}\right)$

(b)  For complementary function, $c'' - 2c' - 8c = 0$, TRY $c = A e^{mx}$

$$\therefore \; m^2 - 2m - 8 = (m-4)(m+2) = 0$$

$$\therefore \; m = 4 \;\text{ or }\; m = -2 \;\;\Rightarrow\;\; c = A e^{-2x} + B e^{4x}$$

For particular integral, TRY $p \;=\; Dx^2 + Ex + F + Gx e^{-2x}$

$$\therefore \; p' \;=\; 2Dx + E - 2Gx e^{-2x} + G e^{-2x}$$

$$\therefore \; p'' \;=\; 2D + 4Gx e^{-2x} - 4G e^{-2x}$$

But $p'' - 2p' - 8p = x^2 + e^{-2x}$

$$\Rightarrow \;\; \underbrace{-8Dx^2}_{1} - \underbrace{(4D + 8E)x}_{0} + \underbrace{2D - 2E - 8F}_{0} - \underbrace{6G e^{-2x}}_{1} = x^2 + e^{-2x}$$

$$\left.\begin{array}{r} 8D = -1 \\ D + 2E = 0 \\ D - E - 4F = 0 \\ 6G = -1 \end{array}\right\} \quad \begin{array}{l} D = -1/8 \\ E = 1/16 \\ F = -3/64 \\ G = -1/6 \end{array}$$

Therefore general solution is

$$y = A\,e^{-2x} + B\,e^{4x} - \tfrac{1}{64}\left(8x^2 - 4x + 3\right) - \tfrac{1}{6}\,x\,e^{-2x}$$

This is an example where the most obvious thing we might have tried for the particular integral, $p = Dx^2 + Ex + F + He^{-2x}$, would not have worked because $e^{-2x}$ is part of the complementary function. Replacing that exponential with an $x\,e^{-2x}$ term is the simplest reasonable amendment, but it could have been that an $x^2 e^{-2x}$ contribution was also required. If we had included all three of these terms in the particular integral, $Gx e^{-2x} + He^{-2x} + Lx^2 e^{-2x}$, we would have found that $G = -1/6$, $L = 0$ and $H$ was undetermined.

(c)   For complementary function, $c'' + c' + c = 0$,   TRY   $c = A\,e^{mx}$

$$\therefore \quad m^2 + m + 1 = 0 \quad \Rightarrow \quad m = \tfrac{-1 \pm \sqrt{1-4}}{2} = -\tfrac{1}{2} \pm i\tfrac{\sqrt{3}}{2}$$

$\left[A\,e^{+i\omega_0 x} + B\,e^{-i\omega_0 x}\right]e^{-x/2}$

$$\therefore \quad c = \left[D\sin(\omega_0 x) + E\cos(\omega_0 x)\right]e^{-x/2} \quad \text{where } \omega_0 = \tfrac{\sqrt{3}}{2}$$

For particular integral,   TRY   $p = \quad F\sin(\omega x) + \quad G\cos(\omega x)$

$$\therefore \quad p' = \quad \omega F\cos(\omega x) - \quad \omega G\sin(\omega x)$$

$$\therefore \quad p'' = -\omega^2 F\sin(\omega x) - \omega^2 G\cos(\omega x)$$

But   $p'' + p' + p = \cos(\omega x)$

$$\Rightarrow \quad \underbrace{\left[F - \omega G - \omega^2 F\right]}_{0}\sin(\omega x) + \underbrace{\left[G + \omega F - \omega^2 G\right]}_{1}\cos(\omega x) = \cos(\omega x)$$

Such equations are most conveniently solved using a matrix method.

$$\begin{pmatrix} 1-\omega^2 & -\omega \\ \omega & 1-\omega^2 \end{pmatrix}\begin{pmatrix} F \\ G \end{pmatrix} = \begin{pmatrix} 0 \\ 1 \end{pmatrix}$$

$$\Rightarrow \quad \begin{pmatrix} F \\ G \end{pmatrix} = \begin{pmatrix} 1-\omega^2 & -\omega \\ \omega & 1-\omega^2 \end{pmatrix}^{-1}\begin{pmatrix} 0 \\ 1 \end{pmatrix} = \frac{1}{\omega^4 - \omega^2 + 1}\begin{pmatrix} \omega \\ 1-\omega^2 \end{pmatrix}$$

$\omega_0 = \tfrac{\sqrt{3}}{2}$

$$\therefore \quad y = \left[D\sin(\omega_0 x) + E\cos(\omega_0 x)\right]e^{-x/2}$$

$$+ \frac{\omega\sin(\omega x) + (1-\omega^2)\cos(\omega x)}{\omega^4 - \omega^2 + 1}$$

This solution describes the motion of a *driven damped oscillator*. The term on the right-hand side of the original ODE

$$\frac{d^2y}{dx^2} + \frac{dy}{dx} + y = \cos(\omega x)$$

represents the driving force, $y$ is the displacement of the system, and $x$ is time. The effect of the (positive $dy/dx$) damping term is to make the complementary function $c(x)$ tend to zero as $x \to \infty$; $c(x)$ is the *transient* solution, and $p(x)$ is the *steady-state* solution which describes the motion once the transient has decayed away.

An alternative, and usually easier, route to the steady-state solution makes use of the material in chapter 7:

TRY  $p = \mathcal{R}e\{z\}$   where  $z = z_0\,e^{i\omega x}$

with the constant $z_0$ being a complex number. The useful properties discussed in section 7.8 mean that $z$ satisfies the ODE

$$\frac{d^2z}{dx^2} + \frac{dz}{dx} + z = e^{i\omega x}$$

The substitution of the trial $z(x)$ shows it to be a viable solution as long as

$$[-\omega^2 + i\omega + 1]z_0 = 1$$

Hence

$$z_0 = \frac{1}{1 - \omega^2 + i\omega} = \frac{1 - \omega^2 - i\omega}{(1 - \omega^2)^2 + \omega^2} = \frac{(1 - \omega^2) + i(-\omega)}{\omega^4 - \omega^2 + 1}$$

and $\mathcal{R}e\{z\}$ yields the same particular integral as earlier.

The modulus and argument form

$$z_0 = R\,e^{i\phi} \quad \text{where} \quad R = \frac{1}{\sqrt{\omega^4 - \omega^2 + 1}} \quad \text{and} \quad \phi = -\tan^{-1}\left(\frac{\omega}{1 - \omega^2}\right)$$

readily gives the steady-state solution as

$$p = \mathcal{R}e\{z_0\,e^{i\omega x}\} = \mathcal{R}e\{R\,e^{i(\omega x + \phi)}\} = R\cos(\omega x + \phi)$$

so that $R$ is seen to be the amplitude of the response of the damped oscillator, when driven at frequency $\omega$, and $\phi$ is the phase difference between the driving force (input) and the response (output).

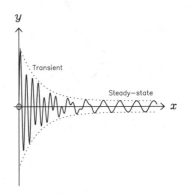

$p = -\beta\sin(\omega x) + \alpha\cos(\omega x)$

where $z_0 = \alpha + i\beta$

$$\frac{d}{dx}[e^{i\omega x}] = i\omega\,e^{i\omega x}$$

$R^2 = z_0^*z_0 = \alpha^2 + \beta^2$

$\tan\phi = \beta/\alpha$

---

**(6)** Solve

$$x^2\frac{d^2y}{dx^2} + 3x\frac{dy}{dx} + y = 0$$

by (i) using the trial solution $y = A\,x^\lambda$, and (ii) with the substitution $u = \ln x$.

(i)  TRY  $y = A x^{\lambda}$  $\Rightarrow$  $\dfrac{dy}{dx} = A \lambda x^{\lambda-1}$  and  $\dfrac{d^2 y}{dx^2} = A \lambda (\lambda - 1) x^{\lambda-2}$

$\therefore\ A x^{\lambda} \left[ \lambda (\lambda - 1) + 3\lambda + 1 \right] = 0\ \ \Rightarrow\ \ \lambda^2 + 2\lambda + 1 = (\lambda + 1)^2 = 0$

The repeated root, $\lambda = -1$, produces just one solution:   $\underline{y = A x^{-1}}$

Since we know that the general solution of a second-order ODE must have two arbitrary constants, we are missing a solution that has a different functional form. Trying $x$ times the first solution, as normally works for the standard $e^{\lambda x}$ case, is not an option since this returns a trial function of the type already tried. The substitution method below resolves the problem.

(ii)  Putting  $u = \ln x$,   $\dfrac{d}{dx} [y] = \dfrac{dy}{dx} = \underbrace{\dfrac{du}{dx}}_{1/x} \dfrac{dy}{du} = \dfrac{1}{x} \dfrac{d}{du} [y]$

$\therefore\ \ 3x \dfrac{dy}{dx} = 3x \dfrac{1}{x} \dfrac{dy}{du} = 3 \dfrac{dy}{du}$   and

$x^2 \dfrac{d^2 y}{dx^2} = x^2 \dfrac{d}{dx} \left[ \dfrac{1}{x} \dfrac{dy}{du} \right] = x \underbrace{\dfrac{d}{dx} \left[ \dfrac{dy}{du} \right]}_{d/du} - \dfrac{dy}{du} = \dfrac{d^2 y}{du^2} - \dfrac{dy}{du}$

$\Rightarrow\ \ \dfrac{d^2 y}{du^2} + 2 \dfrac{dy}{du} + y = 0$

TRY  $y = A e^{mu}$    $\Rightarrow$    $m^2 + 2m + 1 = (m + 1)^2 = 0$

So again we get a repeated root but, since it is the standard homogeneous case, we know that $u e^{mu}$ now provides a second solution. Hence, with $e^u = x$,

$$y = A e^{-u} + B u e^{-u} = \underline{(A + B \ln x) x^{-1}}$$

# 14 Partial differential equations

## 14.1 Elementary cases

The previous chapter, on ordinary differential equations, was restricted to functions of a single variable; now let's extend the analysis to multi-parameter problems. As we learnt in chapter 10, such a generalization entails the replacement of ordinary derivatives (like $dy/dx$) with partial ones (such as $\partial y/\partial x$). Hence, the material in this chapter goes under the heading of *partial differential equations*; or PDEs for short. For clarity, most of the examples will involve functions of just two variables.

Let us begin with the simplest PDE of all

$$\left(\frac{\partial z}{\partial x}\right)_y = 0 \qquad (14.1)$$

If a derivative was zero, we would normally deduce that its integral was equal to a constant. In eqn (14.1), $y$ is held fixed; and so any function of it, say $g(y)$, is equivalent to a constant. Hence, the most general solution is

$$z(x,y) = g(y)$$

This elementary example highlights that PDEs give rise to 'functions of integration', in contrast to ODEs which have 'constants of integration'.

Now let's consider a more interesting case. Suppose that we have the exact differential $df = \left[(x-y)/x^2\right] dx + \left[1/x\right] dy$; what is $f(x,y)$? Well, eqn (11.5) shows that

$$\left(\frac{\partial f}{\partial x}\right)_y = \frac{x-y}{x^2} \quad \text{and} \quad \left(\frac{\partial f}{\partial y}\right)_x = \frac{1}{x} \qquad (14.2)$$

Starting with the second expression, because it's less complicated, it states that the derivative of $f(x,y)$ with respect to $y$, with $x$ treated like a constant, is $1/x$. Thinking in integral terms, under the same conditions, we can deduce that

$$f(x,y) = \frac{y}{x} + g(x) \qquad (14.3)$$

The best way to get a handle on $g(x)$ is to differentiate eqn (14.3) with respect to $x$, with $y$ held fixed, and compare the result with the first part of eqn (14.2):

$$\left(\frac{\partial f}{\partial x}\right)_y = \frac{dg}{dx} - \frac{y}{x^2} = \frac{1}{x} - \frac{y}{x^2}$$

where we have replaced $(\partial g/\partial x)_y$ with $dg/dx$ since, by definition, $g(x)$ is a function of $x$ only. Thus $dg/dx = 1/x$, and so $g(x) = \ln x + K$ where $K$ is a true constant. Hence, the general solution to our exact differential is

$$f(x,y) = \frac{y}{x} + \ln x + K$$

which can, and should, be checked by making sure that it returns eqn (14.2).

For the final illustration in this section, let's consider an easy second-order example: $\partial^2 z/\partial x\,\partial y = 0$. This can be tackled by remembering what the mixed partial derivative stands for

$$\frac{\partial^2 z}{\partial x\,\partial y} = \frac{\partial}{\partial x}_y\left(\frac{\partial z}{\partial y}\right)_x = 0 \qquad (14.4)$$

Integrating eqn (14.4) with respect to $x$, while treating $y$ like a constant, yields

$$\left(\frac{\partial z}{\partial y}\right)_x = g(y)$$

A second integration, with respect to $y$ at fixed $x$, gives the general solution as

$$z(x,y) = G(y) + h(x) \qquad (14.5)$$

where $G(y) = \int g(y)\,dy$ is also a function of $y$ only. Although eqn (14.5) may seem uninformative, because $G(y)$ and $h(x)$ cannot be determined without suitable boundary conditions, it does tell us that $z(x,y)$ is the sum of separate functions of $x$ and $y$; that is to say, there can be no mixed terms.

All the PDEs in this section were elementary, in that they could be solved by direct integration as long as we thought carefully about what the partial derivatives actually meant. Now let's turn to a more realistic situation.

## 14.2 Separation of variables

Most PDEs of physical interest, such as the *wave equation*, tend to be of second order

$$\frac{\partial^2 y}{\partial x^2} = \frac{1}{c^2}\frac{\partial^2 y}{\partial t^2} \qquad (14.6)$$

where $y$ could be the 'vertical' displacement of a string, $x$ the position along it, $t$ is time, and $c$ the speed of the wave. There are several different ways of solving PDEs, but almost all of them are beyond the scope of this book. We will be focusing on just one particular method called the *separation of variables*; any similarity in the name with the approach described in section 13.2 is incidental, as they are quite unrelated.

Before starting the main part of this section, however, we should mention an interesting point about eqn (14.6). The substitution of $u = x + ct$ and $v = x - ct$ transforms the wave equation into the form met in eqn (14.4), $\partial^2 y / \partial u\, \partial v = 0$; this was shown in example (5) of section 10.7. Hence, the general solution of eqn (14.6) can be written as

$$y(x, t) = G(x + ct) + h(x - ct)$$

where the functions G and h must be determined from suitable boundary conditions. This is the basis of *d'Alembert's formula*.

Returning to the method of the separation of variables, this is best explained with reference to a specific example. The one we will use can be stated in terms of the following question: 'A function $f(x, y)$ is defined in the strip $x \geqslant 0$ and $0 \leqslant y \leqslant a$, and obeys *Laplace's equation*

$$\frac{\partial^2 f}{\partial x^2} + \frac{\partial^2 f}{\partial y^2} = 0 \tag{14.7}$$

If $f(x, y) \to 0$ as $x \to \infty$, and satisfies the boundary conditions

$$f(x, 0) = f(x, a) = 0$$

and $\quad f(0, y) = \sin\left(\frac{\pi y}{a}\right) + 2 \sin\left(\frac{2\pi y}{a}\right) \tag{14.8}$

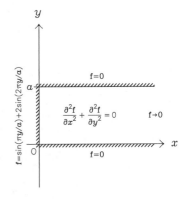

find the solution of Laplace's equation.'

As a useful preliminary to tackling this sort of problem, it is often helpful to sketch a diagram which highlights the region of interest, and shows the relevant boundary conditions. The first step in the separation of variables, and the thing that gives the method its name, is to try a solution of the form

$$f(x, y) = X(x)\, Y(y) \tag{14.9}$$

We should emphasize that this simply represents an educated guess as to what the solutions of eqn (14.7) might look like, a product of independent functions of $x$ and $y$ (X and Y respectively), but we're not sure that it is correct; hence eqn (14.9) is prefaced with the word 'try' rather than 'let' or 'put'.

To see whether eqn (14.9) works, we need to substitute it into eqn (14.7) and check that viable solutions emerge:

$$\left(\frac{\partial f}{\partial x}\right)_y = Y(y)\left(\frac{\partial X}{\partial x}\right)_y = Y\frac{\mathrm{d}X}{\mathrm{d}x}$$

$$Y\frac{\mathrm{d}^2 X}{\mathrm{d}x^2} + X\frac{\mathrm{d}^2 Y}{\mathrm{d}y^2} = 0$$

$$\left(\frac{\partial f}{\partial y}\right)_x = X(x)\left(\frac{\partial Y}{\partial y}\right)_x = X\frac{\mathrm{d}Y}{\mathrm{d}y}$$

where we have replaced the partial derivatives with ordinary ones because, by the definition of eqn (14.9), X and Y are respectively functions of $x$ and $y$ only. Dividing the expression above by XY, and rearranging the result, we obtain

$$\frac{1}{X}\frac{\mathrm{d}^2 X}{\mathrm{d}x^2} = -\frac{1}{Y}\frac{\mathrm{d}^2 Y}{\mathrm{d}y^2} \tag{14.10}$$

At this stage, there is a simple but subtle point that requires some thought: the left- and right-hand sides of eqn (14.10) are separately functions of $x$ and $y$, and yet are equal to each other. This can only happen if neither is actually a function of $x$ or $y$, but are equal to a constant. Calling the latter $\omega^2$, for reasons that will become obvious later, our original PDE of eqn (14.7) is equivalent to the two simultaneous ODEs

$$\frac{d^2 X}{dx^2} = \omega^2 X \quad \text{and} \quad \frac{d^2 Y}{dy^2} = -\omega^2 Y \quad (14.11)$$

These can either be solved by using the material in section 13.5, that is by trying $X = A e^{mx}$ and so on, or by recognizing the relationship between eqn (14.11) and simple harmonic motion. In any case, eqns (14.9) and (14.11) combine to give the form of the separable solutions of eqn (14.7) as

$$f(x, y) = \left[ A e^{\omega x} + B e^{-\omega x} \right] \left[ C \cos(\omega y) + D \sin(\omega y) \right] \quad (14.12)$$

where $A$, $B$, $C$, and $D$ are constants of integration.

Having dealt with the PDE per se, the remainder of our task consists of applying the boundary conditions. Starting with $f(x, y) \rightarrow 0$ as $x \rightarrow \infty$, this can only

$$A_n = 0$$

$$C_n = 0$$

$$\omega_n = \frac{n\pi}{a}$$

$$f_n(x, y) \propto \sin(\omega_n y)\, e^{-\omega_n x}$$

be satisfied if $A = 0$ (making the natural assignment that $\omega$ is real and positive). Similarly, $f(x, 0) = 0$ requires that $C = 0$ because the alternative of $B = 0$ would lead to the unacceptable conclusion that $f = 0$ everywhere. Next, $f(x, a) = 0$ means that $\sin(\omega a) = 0$; hence $\omega a = n\pi$ where $n = 0, 1, 2, 3, \ldots$. The fact that $n$, and in turn $\omega$, can take on a multitude of values shows that there are a large number of solutions of the form of eqn (14.12). Labelling each with the suffix $n$, the general solution of eqn (14.7) is given by their sum

$$f(x, y) = \sum_{n=1}^{\infty} E_n \sin\left(\frac{n\pi y}{a}\right) e^{-n\pi x/a} \quad (14.13)$$

where we have relabelled $(BD)_n$ as $E_n$, and set $A_n = C_n = 0$. The reason for the summation is that for linear PDEs, such as eqn (14.7), a linear combination of solutions is also a solution. Finally, by comparing $f(0, y)$ with the expression in the boundary condition of eqn (14.8), we find that $E_1 = 1$, $E_2 = 2$, and all the other $E_n$ coefficients are zero. Hence, the solution is

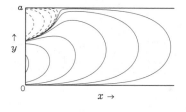

$$f(x, y) = \sin\left(\frac{\pi y}{a}\right) e^{-\pi x/a} + 2 \sin\left(\frac{2\pi y}{a}\right) e^{-2\pi x/a} \quad (14.14)$$

Strictly speaking, we should now check that eqn (14.14) satisfies both the PDE of eqn (14.7) and the requirements of eqn (14.8).

Although we have illustrated the method of the separation of variables with a specific example, the steps for dealing with other problems are virtually identical. The power of the technique lies in its ability to reduce a PDE to an equivalent set of ODEs. On a practical note, the choice of the constant used at the stage of eqns (14.10) and (14.11), and the subsequent order in which the boundary conditions are implemented, plays an important part in facilitating a successful outcome. In this context, the recommended preliminary diagram is helpful because it suggests that the solution of the PDE will be a decaying function in the $x$ direction and an oscillatory one in $y$. The use of a constant other than $\omega^2$, such as $\omega$ or $-\omega^2$, would have led to messier algebra (involving square roots and the imaginary i); if required, the $\omega = 0$ solution has to be considered separately. It is also best to apply the simplest boundary conditions first, especially if they eliminate some of the terms at the stage of eqn (14.12), and leave the most complicated till the end.

## 14.3 Common physical examples

We have already met two PDEs of physical interest in section 14.2; namely, the wave equation of eqn (14.6) and Laplace's equation in eqn (14.7). While the former needs little introduction, the latter, and its more general form in the shape of *Poisson's equation* (where the right-hand side is not zero), is of importance in electrostatics, hydrodynamics, and so on. A third common case is the *diffusion equation*

$$K \frac{\partial^2 u}{\partial x^2} = \frac{\partial u}{\partial t} \tag{14.15}$$

which describes the flow of heat in a metal, the spread of a solute in a solvent, and the like.

The one-dimensional wave equation of eqn (14.6) can be extended to three dimensions by replacing the differential operator $\partial^2/\partial x^2$ with the *Laplacian* $\nabla^2$

$$\nabla^2 \psi = \frac{1}{c^2} \frac{\partial^2 \psi}{\partial t^2} \tag{14.16}$$

where

$$\nabla^2 \psi = \frac{\partial^2 \psi}{\partial x^2} + \frac{\partial^2 \psi}{\partial y^2} + \frac{\partial^2 \psi}{\partial z^2} \tag{14.17}$$

in Cartesian coordinates. Similarly,

$$K \nabla^2 u = \frac{\partial u}{\partial t} \quad \text{and} \quad \nabla^2 f = 0 \tag{14.18}$$

for the diffusion and Laplace's equations. In solving eqn (14.16) with the separation of variables, we now try $\psi(x, y, z, t) = X(x)\,Y(y)\,Z(z)\,T(t)$. If the problem being tackled has an inherent spherical symmetry, such as the hydrogen atom, then, as noted in section 12.3, it is better to work in spherical polar coordinates. With $\nabla^2 \psi$ expressed in terms of $r$, $\theta$, and $\phi$ (which is more complicated than in Cartesians), we try $\psi(r, \theta, \phi, t) = R(r)\,\Theta(\theta)\,\Phi(\phi)\,T(t)$. The radial part of the solution turns out to be related to *Bessel functions*, and the combined angular contribution is closely linked with *spherical harmonics*. With the application of suitable boundary conditions, a range of discrete possibilities emerges as in eqn (14.13); the suffices used to denote the alternatives have a direct correspondence with the various *quantum numbers* that appear in atomic physics.

$$\nabla^2 \psi = \frac{1}{r^2} \frac{\partial}{\partial r}\left(r^2 \frac{\partial \psi}{\partial r}\right)$$

$$+ \frac{1}{r^2 \sin\theta} \frac{\partial}{\partial \theta}\left(\sin\theta \frac{\partial \psi}{\partial \theta}\right)$$

$$+ \frac{1}{r^2 \sin^2\theta} \frac{\partial^2 \psi}{\partial \phi^2}$$

Finally, as a matter of passing interest, we note that *Schrödinger's equation* (time-independent), which is really a wave equation for 'standing waves', is also an eigenvalue equation

$$\mathcal{H}\psi = \left[ V - \frac{\hbar^2}{2m}\nabla^2 \right]\psi = E\psi \qquad (14.19)$$

$$\underbrace{\phantom{\left[ V - \frac{\hbar^2}{2m}\nabla^2 \right]}}_{\mathcal{H}}$$

where $\mathcal{H}$ is the *Hamiltonian* operator and $E$ is a constant. The wavefunctions, $\psi$, which satisfy eqn (14.19) are called *eigenfunctions* and represent the stationary states of the system; the corresponding values of $E$, which are the eigenvalues, give the related energy levels. Essentially, eqn (14.19) is the continuum form of the matrix definition of eqn (9.17).

## 14.4 Worked examples

> **(1)** For the exact differential $df = y\cos(xy)\,dx + \left[ x\cos(xy) + 2y \right] dy$ find $f(x, y)$.

$$\text{If } f = f(x, y), \quad \text{then} \quad df = \left( \frac{\partial f}{\partial x} \right)_y dx + \left( \frac{\partial f}{\partial y} \right)_x dy$$

$$\therefore \quad \text{Exact} \Rightarrow \qquad \left( \frac{\partial f}{\partial x} \right)_y = y\cos(xy) \qquad - (1)$$

$$\text{and} \quad \left( \frac{\partial f}{\partial y} \right)_x = x\cos(xy) + 2y \quad - (2)$$

Integrating (1) with respect to $x$, at constant $y$, gives

$$f(x, y) = \sin(xy) + g(y) \qquad - (3)$$

Substituting (3) into (2), we obtain

$$\left( \frac{\partial f}{\partial y} \right)_x = x\cos(xy) + \frac{dg}{dy} = x\cos(xy) + 2y$$

$$\therefore \quad \frac{dg}{dy} = 2y \quad \Rightarrow \quad g(y) = y^2 + K$$

$$\text{i.e.} \quad \underline{f(x, y) = \sin(xy) + y^2 + K}$$

$(2) \Rightarrow$

$$f(x, y) = \sin(xy) + y^2 + h(x)$$

$$\therefore \ (1) \Rightarrow \quad \frac{dh}{dx} = 0$$

We could have done this question in a number of different ways. By integrating (2) with respect to $y$ while treating $x$ like a constant, for example, and substituting the resultant formula for $f(x, y)$ into (1) to yield the related function of integration $h(x)$. Alternatively, we could have got to the general solution by directly comparing the two formulae for $f(x, y)$ that result from the integration of (1) and (2).

**(2)** Verify that $u(x, t) = \dfrac{1}{\sqrt{4kt}} \exp\left(-\dfrac{x^2}{4kt}\right)$

is a solution of the diffusion equation $k\dfrac{\partial^2 u}{\partial x^2} = \dfrac{\partial u}{\partial t}$.

$$\left(\frac{\partial u}{\partial t}\right)_x = \exp\left(-\frac{x^2}{4kt}\right)\left[\frac{\partial}{\partial t_x}\left(\frac{1}{\sqrt{4kt}}\right) + \frac{1}{\sqrt{4kt}}\frac{\partial}{\partial t_x}\left(-\frac{x^2}{4kt}\right)\right]$$

$$= \exp\left(-\frac{x^2}{4kt}\right)\left[-\frac{2k}{(4kt)^{3/2}} + \frac{x^2}{4kt^2\sqrt{4kt}}\right]$$

$$= \exp\left(-\frac{x^2}{4kt}\right)\left[\frac{x^2}{t} - 2k\right](4kt)^{-3/2}$$

$$\left(\frac{\partial u}{\partial x}\right)_t = \frac{1}{\sqrt{4kt}}\exp\left(-\frac{x^2}{4kt}\right)\frac{\partial}{\partial x_t}\left(-\frac{x^2}{4kt}\right)$$

$$= -2x\exp\left(-\frac{x^2}{4kt}\right)(4kt)^{-3/2}$$

$$\therefore\ k\frac{\partial^2 u}{\partial x^2} = k\frac{\partial}{\partial x_t}\left(\frac{\partial u}{\partial x}\right)_t$$

$$= -2k\exp\left(-\frac{x^2}{4kt}\right)\left[\frac{\partial}{\partial x_t}(x) + x\frac{\partial}{\partial x_t}\left(-\frac{x^2}{4kt}\right)\right](4kt)^{-3/2}$$

$$\therefore\ \underline{k\frac{\partial^2 u}{\partial x^2}} = -2k\exp\left(-\frac{x^2}{4kt}\right)\left[1 - \frac{x^2}{2kt}\right](4kt)^{-3/2} = \underline{\left(\frac{\partial u}{\partial t}\right)_x}$$

This solution of the diffusion equation shows how heat (as in temperature) in a metal bar, or the concentration of a solvent in solute, spreads out with time from an initial point source. For a given time $t$, the spatial distribution (in $x$) takes the form of the ubiquitous Gaussian function (met in probability and statistics as the *normal* distribution). When centred at the origin, its shape is characterized by the quadratic exponential $\exp\left(-\frac{x^2}{2\sigma^2}\right)$ where the width is given by the constant $\sigma$ (technically known as the 'standard deviation', or $\sigma^2$ the 'variance'). In the diffusion case $\sigma \propto \sqrt{t}$, so that the spread doubles when the time quadruples and so on. Incidentally, the factor of $1/\sqrt{4kt}$ outside the exponential is a normalization term which ensures that the integral of the distribution remains constant with time; in other words, the total number of solute particles (say) is fixed even though they are spreading out. In the standard Gaussian distribution, $\exp\left(-\frac{x^2}{2\sigma^2}\right)$, this normalization prefactor is $\left(\sigma\sqrt{2\pi}\right)^{-1}$.

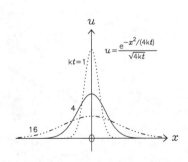

**(3)** Given the Schrödinger equation for free electrons in two dimensions

$$\frac{\partial^2 \psi}{\partial x^2} + \frac{\partial^2 \psi}{\partial y^2} + \frac{8\pi^2 m E \psi}{h^2} = 0$$

obtain the wavefunctions $\psi(x,y)$, and the energy levels $E$, subject to the boundary conditions $\psi = 0$ at $x=0$, $x=a$, $y=0$, and $y=b$.

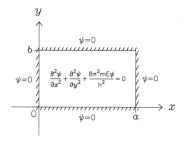

TRY   $\psi(x,y) = X(x)\,Y(y)$

$$\therefore \quad Y\frac{d^2 X}{dx^2} + X\frac{d^2 Y}{dy^2} + \frac{8\pi^2 m E X Y}{h^2} = 0$$

$$\therefore \quad \frac{1}{X}\frac{d^2 X}{dx^2} + \frac{8\pi^2 m E}{h^2} = -\frac{1}{Y}\frac{d^2 Y}{dy^2} = \omega^2 \quad \text{(a constant)}$$

i.e.   $\dfrac{d^2 X}{dx^2} = -\underbrace{\left(\dfrac{8\pi^2 m E}{h^2} - \omega^2\right)}_{\Omega^2} X$   and   $\dfrac{d^2 Y}{dy^2} = -\omega^2 Y$

$\therefore$   Separable solutions are of the form

$$\psi(x,y) = \big[A\sin(\Omega x) + B\cos(\Omega x)\big]\big[C\sin(\omega y) + D\cos(\omega y)\big]$$

But   $\psi(0,y) = 0 \ \Rightarrow\ B = 0$

$\psi(x,0) = 0 \ \Rightarrow\ D = 0$

$\psi(a,y) = 0 \ \Rightarrow\ \sin(\Omega a) = 0$   i.e. $\Omega = \dfrac{k\pi}{a}$   for integer $k$

$\psi(x,b) = 0 \ \Rightarrow\ \sin(\omega b) = 0$   i.e. $\omega = \dfrac{l\pi}{b}$   for integer $l$

$\therefore$   $\psi_{kl}(x,y) = A_{kl}\sin\left(\dfrac{k\pi x}{a}\right)\sin\left(\dfrac{l\pi y}{b}\right)$   for $k,l = 1,2,3,\ldots$

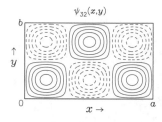

$$\Omega^2 = \frac{8\pi^2 m E}{h^2} - \omega^2 \quad \Rightarrow \quad E_{kl} = \frac{h^2}{8m}\left(\frac{k^2}{a^2} + \frac{l^2}{b^2}\right) = \frac{h^2(\Omega^2 + \omega^2)}{8\pi^2 m}$$

This is a two-dimensional version of the standard 'particle in a box' quantum mechanics example. In the one-dimensional case, the problem reduces graphically to one of drawing the 'normal modes' of a string whose ends are tied down and separated by a given length $L$ (say). This leads to the condition that $L$ can only accommodate an integer number of half-wavelengths; that is to say, $L = n\lambda/2$ where $n = 1, 2, 3, \ldots$ and $\lambda$ is the wavelength. According to de Broglie, the momentum $p$ associated with a quantum wavelength $\lambda$ is given by $p = h/\lambda$ where $h$ is Planck's constant. Thus the related kinetic energy $p^2/2m$ (where $m$

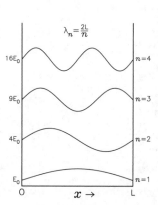

is the mass of the 'particle') is $h^2/(2m\lambda^2) = h^2n^2/(8mL^2)$. Our formula for $E_{kl}$ is just the sum of two one-dimensional contributions, and $\psi_{kl}(x, y)$ is simply the product of the normal mode solutions in the $x$ and $y$ directions.

---

**(4)** Laplace's equation in plane polar coordinates $(r, \theta)$ is

$$\frac{\partial^2 \Phi}{\partial r^2} + \frac{1}{r}\frac{\partial \Phi}{\partial r} + \frac{1}{r^2}\frac{\partial^2 \Phi}{\partial \theta^2} = 0$$

Show, by separating the variables, that there are solutions of the form

$$\Phi(r, \theta) = \left(A_0 \theta + B_0\right)\left(C_0 \ln r + D_0\right) \quad \text{and}$$

$$\Phi(r, \theta) = \left[A_n \cos(n\theta) + B_n \sin(n\theta)\right]\left(C_n r^n + D_n r^{-n}\right)$$

where the $A$, $B$, $C$, $D$, and $n$ are constants. If $\Phi$ is a single-valued function of $\theta$, how does this restrict $n$? Find the solution $\Phi(r, \theta)$ for $0 < r \leqslant a$ when (a) $\Phi(a, \theta) = T \cos \theta$; and (b) $\Phi(a, \theta) = T \cos^3 \theta$.

---

TRY $\quad \Phi(r, \theta) = R(r)\,\Theta(\theta)$

$$\therefore \quad \Theta \frac{d^2 R}{dr^2} + \frac{\Theta}{r}\frac{dR}{dr} + \frac{R}{r^2}\frac{d^2 \Theta}{d\theta^2} = 0$$

$$\therefore \quad \frac{r^2}{R}\frac{d^2 R}{dr^2} + \frac{r}{R}\frac{dR}{dr} = -\frac{1}{\Theta}\frac{d^2 \Theta}{d\theta^2} = n^2 \quad \text{(a constant)}$$

i.e. $\quad r^2 \frac{d^2 R}{dr^2} + r\frac{dR}{dr} = n^2 R \quad$ and $\quad \frac{d^2 \Theta}{d\theta^2} = -n^2 \Theta$

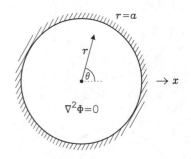

Consider first the special case of $n = 0$

$$r\frac{d^2 R}{dr^2} + \frac{dR}{dr} = 0 \quad \text{and} \quad \frac{d^2 \Theta}{d\theta^2} = 0$$

$$\therefore \quad \frac{d}{dr}\left(r\frac{dR}{dr}\right) = 0 \quad \text{and} \quad \frac{d\Theta}{d\theta} = A_0$$

$$\therefore \quad r\frac{dR}{dr} = C_0 \quad \text{and} \quad \Theta(\theta) = A_0 \theta + B_0$$

But $\quad \int dR = C_0 \int \frac{dr}{r} \quad \Rightarrow \quad R(r) = C_0 \ln r + D_0$

i.e. The $n = 0$ solution is $\quad \underline{\Phi(r, \theta) = \left(A_0 \theta + B_0\right)\left(C_0 \ln r + D_0\right)}$

More generally, if $n \neq 0$,    $\Theta(\theta) = A_n \cos(n\theta) + B_n \sin(n\theta)$

For    $r^2 \dfrac{d^2 R}{dr^2} + r \dfrac{dR}{dr} = n^2 R$,    TRY  $R(r) = r^\alpha$

$\therefore\ \ \alpha(\alpha-1)r^\alpha + \alpha r^\alpha = n^2 r^\alpha$    $\Rightarrow$    $\alpha^2 = n^2$

$\therefore\ \ \alpha = \pm n$    $\Rightarrow$    $R(r) = C_n r^n + D_n r^{-n}$

i.e.    $\underline{\Phi(r,\theta) = \left[ A_n \cos(n\theta) + B_n \sin(n\theta) \right]\left( C_n r^n + D_n r^{-n} \right)}$

If $\Phi$ is a single-valued function of $\theta$, then  $\Phi(r,\theta) = \Phi(r, \theta + 2\pi k)$ for any integer $k$. This can be satisfied by the $n \neq 0$ solutions as long as $n$ is an integer, but not by the $n=0$ case:  $\underline{n = 1, 2, 3, \dots}$

For the solutions to be finite at $r=0$, $D_n=0$. Therefore, with $C_n=1$, the general solution is

$$\Phi(r,\theta) = \sum_{n=1}^{\infty} \left[ A_n \cos(n\theta) + B_n \sin(n\theta) \right] r^n$$

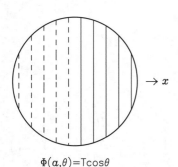

$\Phi(a,\theta)=T\cos\theta$

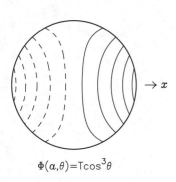

$\Phi(a,\theta)=T\cos^3\theta$

(a)   $\Phi(a,\theta) = T\cos\theta$    $\Rightarrow$    $A_n=0$ if $n \neq 1$   and   $B_n=0$ for all $n$

$$a A_1 = T \quad \Rightarrow \quad \underline{\Phi(r,\theta) = \dfrac{Tr\cos\theta}{a}}$$

(b)   $\cos^3\theta = \left( \dfrac{e^{i\theta} + e^{-i\theta}}{2} \right)^3 = \dfrac{e^{i3\theta} + e^{-i3\theta} + 3\left( e^{i\theta} + e^{-i\theta} \right)}{8}$

$\therefore\ \ \Phi(a,\theta) = T\cos^3\theta = \dfrac{T}{4}\left( \cos 3\theta + 3\cos\theta \right)$    $\Rightarrow$    $B_n=0$ for all $n$

$$a A_1 = \dfrac{3T}{4}, \quad a^3 A_3 = \dfrac{T}{4} \quad \text{and} \quad A_n=0 \text{ otherwise}$$

$\therefore\ \ \underline{\Phi(r,\theta) = \dfrac{3Tr\cos\theta}{4a} + \dfrac{Tr^3\cos 3\theta}{4a^3}}$

# 15 Fourier series and transforms

## 15.1 Approximating periodic functions

In chapter 6, we saw how a Taylor series could be used to mimic an arbitrary function by a simple low-order polynomial in the neighbourhood of a particular point. An alternative approximation is provided by a Fourier series, which is appropriate when the curve of interest is periodic

$$f(x) = f(x + \lambda)$$

$$
\begin{aligned}
f(x) \approx \tfrac{1}{2} a_0 &+ a_1 \cos(kx) + a_2 \cos(2kx) + a_3 \cos(3kx) + \cdots \\
&+ b_1 \sin(kx) + b_2 \sin(2kx) + b_3 \sin(3kx) + \cdots
\end{aligned}
\tag{15.1}
$$

$$k = \frac{2\pi}{\lambda}$$

where the $a$ and $b$ coefficients, as well as k, are constants. In fact, eqn (15.1) is the counterpart of eqn (6.1). Any repeating function can be represented as a sum of sine and cosine *harmonics*. We could replace the $a\cos(kx) + b\sin(kx)$ with $A\cos(kx + \phi)$ or $A\sin(kx + \phi)$, due to their trigonometric equivalence, but the former is usually preferable because it's linear. The period of eqn (15.1), $\lambda$, is given by the lowest harmonic as $\lambda = 2\pi/k$.

Multiplying eqn (15.1) through by $\cos(nkx)$, and $\sin(nkx)$, and integrating both sides over one repeat interval, it can be shown that the Fourier coefficients are given by

$$
a_n = \frac{2}{\lambda} \int_0^\lambda f(x) \cos(nkx)\, dx \quad \text{and} \quad b_n = \frac{2}{\lambda} \int_0^\lambda f(x) \sin(nkx)\, dx \tag{15.2}
$$

where $n = 1, 2, 3, \ldots$, and relies on the fact that the sine and cosine harmonics are orthogonal. This is to say the integral of $\sin(mkx)\sin(nkx)$ and $\cos(mkx)\cos(nkx)$ from $x = 0$ to $x = \lambda$ is zero if $m \neq n$; and is nought over this range for any mixed combination of sines and cosines. The reason for the odd-looking factor of a half with the $a_0$ in eqn (15.1) is that the formulae of eqn (15.2) then work even for $n = 0$; the corresponding term with $b_0$ has been omitted because $\sin(0) = 0$.

$$\int_0^\lambda \sin(mkx)\cos(nkx)\, dx = 0$$

Functions that are not periodic can also be approximated by a Fourier series, as long as we are only interested in $f(x)$ over a limited range: $0 \leqslant x \leqslant L$, say. Then we can make it pseudo-repeating by arbitrarily defining $f(x) = f(x + L)$, or by letting $f(x) = f(-x)$ and $f(x) = f(x + 2L)$, or by setting $f(x) = -f(-x)$ and $f(x) = f(x + 2L)$, and so on; $k = 2\pi/L$ if the period is $L$, and $k = \pi/L$ if it's $2L$.

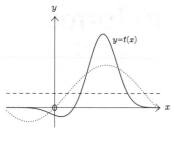

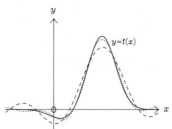

While each possibility will lead to a different Fourier expansion, there being no sine terms if $f(x) = f(-x)$ and no cosine terms if $f(x) = -f(-x)$, they will all give a reasonable approximation in the important region $0 \leqslant x \leqslant L$.

## 15.2 Taylor versus Fourier series

While a Taylor series provides a very good representation of $y = f(x)$ near the expansion point $x = x_0$, it soon yields a poor one as $|x - x_0|$ becomes large. By comparison, a Fourier series is not expected to do an accurate job anywhere in particular but mimics the overall shape of the curve well. The general characteristics of the latter, with its high-frequency oscillatory behaviour when sharp features are involved, means that a term-by-term differentiation is an extremely ill-advised operation; a corresponding integration is much safer, as the wiggles tend to cancel out. At a more technical level, a Taylor series requires that $f(x)$ be differentiable many times over whereas the function, and its gradient $f'(x)$, need only be piecewise continuous for a Fourier series.

## 15.3 Complex Fourier series

A more compact version of eqn (15.1) can be obtained by using the complex number relationship of eqn (7.11), which ties together sines and cosines through the imaginary exponential

$$f(x) = \sum_{n=-\infty}^{\infty} c_n \, e^{inkx} \qquad (15.3)$$

where the coefficients, $c_n$, are complex. If necessary, the $c$ can be related to the $a$ and $b$ in eqn (15.1) with $a_n = (c_n + c_{-n})$ and $b_n = i(c_n - c_{-n})$. With eqn (15.3), there is only one expression for the coefficients

$$c_n = \frac{1}{\lambda} \int_0^{\lambda} f(x) \, e^{-inkx} \, dx \qquad (15.4)$$

Given the periodicity of the integrand above, and those in eqn (15.2), the limits could also be set from $x = -\lambda/2$ to $x = +\lambda/2$. This symmetric choice for the integration period can be analytically advantageous.

## 15.4 Fourier integral

If the repeat interval of the Fourier series is taken to be infinitely long, $f(x)$ is no longer required to be periodic. Carrying out this limiting procedure of $\lambda \to \infty$ (so that $k \to 0$) carefully, we find that the summation of eqn (15.3) becomes an integral

$$f(x) = \int_{-\infty}^{\infty} F(k)\, e^{ikx}\, dk \qquad (15.5)$$

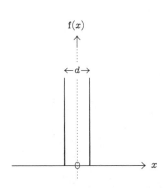

where the 'coefficient function', $F(k)$, is given by

$$F(k) = \frac{1}{2\pi} \int_{-\infty}^{\infty} f(x)\, e^{-ikx}\, dx \qquad (15.6)$$

$F(k)$ is known as the *Fourier transform* of $f(x)$, and $f(x)$ the *inverse* Fourier transform of $F(k)$. Actually, which one we call the transform, and which one the inverse, is just a matter of convention; the important thing is that they come as a related pair, with opposite signs in the exponential. Even the detailed definition, in terms of the location of the $2\pi$, can vary. Sometimes both eqns (15.5) and (15.6) are written with integral pre-factors of $(2\pi)^{-1/2}$, making them look more symmetrical; it's also possible to have the $2\pi$ in both the exponents, as in $\pm i2\pi kx$, and nowhere else.

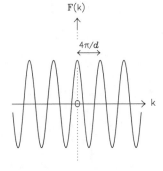

Let us consider a couple of examples. First of all,

$$f(x) = \delta\!\left(x - \tfrac{d}{2}\right) + \delta\!\left(x + \tfrac{d}{2}\right)$$

where the *Dirac delta-function* $\delta(x - x_0)$ is a very narrow spike of unit area at $x = x_0$. Substituting for $f(x)$ into eqn (15.6), and using the property that

$$\int_{-\infty}^{\infty} g(x)\, \delta(x - x_0)\, dx = g(x_0)$$

the Fourier transform of two sharp peaks separated by a distance $d$ is found to be a cosine function with a period that is inversely proportional to $d$:

$$F(k) = \frac{1}{2\pi}\left(e^{ikd/2} + e^{-ikd/2}\right) \propto \cos(kd/2)$$

Similarly, putting

$$f(x) = \begin{cases} 1 & \text{for } -\tfrac{d}{2} \leqslant x \leqslant \tfrac{d}{2} \\ 0 & \text{otherwise} \end{cases}$$

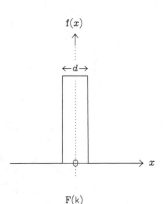

leads to the result that the Fourier transform of a 'top-hat', or 'box', function is related to the *sinc function* $(\sin\theta/\theta)$

$$F(k) = \frac{-1}{i2\pi k}\left[e^{-ikx}\right]_{-d/2}^{d/2} \propto \frac{\sin(kd/2)}{k}$$

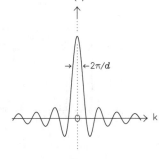

Both these examples indicate that the width of the features in $F(k)$ is inversely proportional to the characteristic spread of the structure in $f(x)$: the narrower the function the broader its Fourier transform, and vice versa.

## 15.5  Some formal properties

Hitherto we have implicitly assumed that $f(x)$ is real, in that $f(x) = f(x)^*$, and can be plotted like an ordinary graph, but there is nothing to stop it from being complex as far as eqns (15.5) and (15.6) are concerned. Indeed, $F(k)$ is complex in general. If $f(x)$ is real, as it usually is, then its Fourier transform can be shown to be *conjugate symmetric*; that is, $F(-k) = F(k)^*$. Other properties of $F(k)$ can also be derived when $f(x)$ exhibits symmetry about the origin: $F(-k) = F(k)$ if $f(x)$ is an even function, and $F(-k) = -F(k)$ when $f(x)$ is odd. These features can be combined in that $F(k)$ is both real and symmetric if $f(x)$ is too; $F(k)$ is imaginary and antisymmetric when $f(x)$ is real and odd.

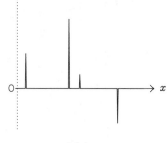

A very important result in Fourier analysis is the *convolution theorem*:

$$f(x) = g(x) \otimes h(x) \quad \Longleftrightarrow \quad F(k) = 2\pi\, G(k) \times H(k) \tag{15.7}$$

where the $\otimes$ denotes a convolution, and $F(k)$, $G(k)$, and $H(k)$ are the Fourier transforms of $f(x)$, $g(x)$, and $h(x)$ respectively. A convolution between two functions is defined by the integral

$$g(x) \otimes h(x) = \int_{-\infty}^{\infty} g(y)\, h(x-y)\, dy = \int_{-\infty}^{\infty} g(x-y)\, h(y)\, dy \tag{15.8}$$

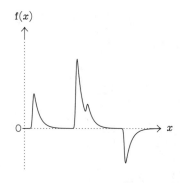

and represents a blurring of $g(x)$ by $h(x)$, or vice versa. If $g(x)$ contains sharp, spiky, structure and $h(x)$ is a broad bell-shaped Gaussian function, for example, then $g(x) \otimes h(x)$ will be a (symmetrically) smeared-out version of $g(x)$. Since it is arbitrary which one of eqns (15.5) and (15.6) is called the Fourier transform and which the inverse, the following also holds

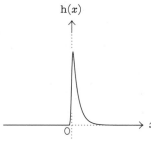

$$f(x) = g(x) \times h(x) \quad \Longleftrightarrow \quad F(k) = \tfrac{1}{2\pi}\, G(k) \otimes H(k) \tag{15.9}$$

Equations (15.7) and (15.9) tells us that a convolution integral in one space is a straightforward product in its Fourier counterpart.

The final property of Fourier transforms that we will consider concerns the *auto-correlation function*, or ACF, which provides information on the distribution of the distances between various parts of the structures in $f(x)$. If there are two spikes with amplitudes $A_1$ and $A_2$ separated by a distance $L$ in $f(x)$, for example, they will contribute a symmetric pair of very sharp components of magnitude $A_1 A_2$ at $x = \pm L$ towards the ACF of $f(x)$. It can also be shown that the ACF of $f(x)$ is related to the Fourier transform of $|F(k)|^2 = F(k)^* F(k)$

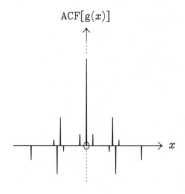

$$\mathrm{ACF}(x) = \int_{-\infty}^{\infty} f(y)^*\, f(x+y)\, dy = 2\pi \int_{-\infty}^{\infty} |F(k)|^2\, e^{ikx}\, dk \tag{15.10}$$

There is always a large positive peak at the origin of the ACF, because everything correlates with itself (separation $= 0$). Incidentally, putting $x = 0$ in eqn (15.10) yields a special case of *Parseval's theorem*

$$\int_{-\infty}^{\infty} |F(k)|^2 \, dk = \frac{1}{2\pi} \int_{-\infty}^{\infty} |f(x)|^2 \, dx \tag{15.11}$$

which is sometimes helpful for evaluating integrals if one side is easier to work out than the other.

## 15.6 Physical examples and insight

Although we began this chapter by thinking about Fourier series as an analogue of the Taylor series for approximating periodic functions, Fourier transforms arise naturally in many branches of science. They occur in both a theoretical context, such as in quantum mechanics, and a physical one, in any experimental situation that entails *diffraction* or *interferometry*. While our discussion has focused on the one-dimensional case, the analysis often needs to be generalized to cater for several parameters. This can be done easily by using vector notation, so that eqn (15.5) becomes

$$f(\mathbf{x}) = \int_{-\infty}^{\infty} \int_{-\infty}^{\infty} \cdots \int_{-\infty}^{\infty} F(\mathbf{k}) \, e^{i\mathbf{k}\cdot\mathbf{x}} \, d\mathbf{k} \tag{15.12}$$

and so on. Equation (15.12) is a surface integral if $\mathbf{x}$ and $\mathbf{k}$ are two-dimensional and a volume one if they are three-dimensional.

To get an intuitive feel for Fourier transforms, let's consider a few examples from *optics* that should be familiar from school physics. We'll begin by describing the formal setup of what is technically known as *Fraunhofer* diffraction: a plane wave of light, of wavelength $\lambda$, strikes a screen which allows some of it through (usually at well-marked slit points) and a pattern of interference *fringes* is seen to be projected onto a distant wall. If $x$ is the distance along the screen, and $q$ is measured across the wall, then it can be shown that the net amplitude of all the waves diffracted in a certain direction, $\psi(q)$, is given by

$$\psi(q) = \psi_0 \int_{-\infty}^{\infty} A(x) \, e^{iqx} \, dx \tag{15.13}$$

where $\psi_0$ is a constant, $A(x)$ is the *aperture function* which specifies how much light is let through by the screen (it's zero for the opaque regions and unity at the locations of the clear slits) and $q$ is related to $\lambda$ and the diffraction angle $\theta$ by $q = 2\pi \sin\theta/\lambda$; $e^{iqx}$ is the part of the plane-wave solution that represents the difference in the path-lengths travelled on going from various points on the screen to a location on the wall. The measured signal, or the number of photons detected, is equal to the intensity $I(q)$

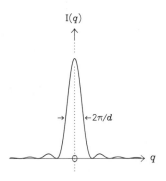

$$I(q) = |\psi(q)|^2 = \psi(q)^* \, \psi(q) \tag{15.14}$$

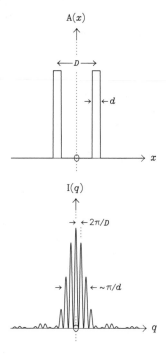

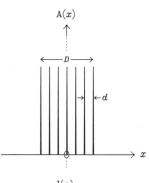

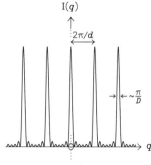

so that the bands of light and dark on the wall are proportional to the modulus-squared of the Fourier transform of the aperture function. Thus, in conjunction with our discussion at the end of section 15.4, the uniform fringes obtained from a *Young's double slit* experiment are mathematically equivalent to

$$I(q) \propto \left[\cos(qd/2)\right]^2 \propto 1 + \cos(qd)$$

Similarly, the diffraction pattern given by a single wide slit is related to

$$I(q) \propto \left[\frac{\sin(qd/2)}{q}\right]^2 \propto \frac{1 - \cos(qd)}{q^2}$$

The diffraction patterns from slightly more complicated cases can often be worked out by analysing the situation as a composite of simpler ones which are linked through the convolution theorem of eqn (15.7). Since a pair of wide slits can be regarded as arising from the convolution of a Young's double slit with a single wide one, for example, the resulting bands of light and dark will consist of the product of a set of uniform (cosine) fringes and a sinc function (squared). Similarly, as a short array of evenly spaced (narrow) slits can be constructed by multiplying an infinitely long 'comb' function and a single wide slit, its diffraction pattern will be made of the convolution of a uniform set of spikes and a sinc function (squared).

Finally, we should note that the measurement of the intensity, $I(q)$ in eqn (15.14), means that the phase of the complex Fourier component, $\psi(q)$ in eqn (15.13), is lost. While the importance of the latter can be difficult to appreciate, it is essential if we are to draw conclusions about the aperture function reliably: $I(q)$ only tells us about the ACF of $A(x)$ in a straightforward manner, as indicated by eqn (15.10), rather than $A(x)$ itself. This translates into such problems as crystallographers have in trying to infer the structure of a protein, say, given the intensities of the *Bragg* spots.

## 15.7 Worked examples

> **(1)** Show that $\sin(mkx)$ and $\cos(nkx)$, where $m$ and $n = 1, 2, 3, \ldots$, are orthogonal over a period $0 \leqslant x \leqslant 2\pi/k$. Hence, derive the formulae for the coefficients of a Fourier series.

$$\int_0^{2\pi/k} \sin(mkx)\cos(nkx)\,\mathrm{d}x = \frac{1}{2}\int_0^{2\pi/k}\left\{\sin[(m+n)kx] + \sin[(m-n)kx]\right\}\mathrm{d}x$$

$$= -\frac{1}{2}\left[\frac{\cos[(m+n)kx]}{(m+n)k} + \frac{\cos[(m-n)kx]}{(m-n)k}\right]_0^{2\pi/k}$$

$$= \frac{1 - \cos[2\pi(m+n)]}{2(m+n)k} + \frac{1 - \cos[2\pi(m-n)]}{2(m-n)k}$$

$$\therefore \int_0^{2\pi/k} \sin(mkx)\cos(nkx)\,dx = 0 \quad \text{for integer } m \text{ and } n$$

Strictly speaking, the above analysis assumes that $m \neq n$ as we have divided by $m - n$. The result still holds for $m = n$, however, as the calculation simplifies at an earlier stage with $\sin\big[(m-n)kx\big] = 0$.

$$\int_0^{2\pi/k} \sin(mkx)\sin(nkx)\,dx = \tfrac{1}{2}\int_0^{2\pi/k}\Big\{\cos\big[(m-n)kx\big] - \cos\big[(m+n)kx\big]\Big\}\,dx$$

$$= \tfrac{1}{2}\left[\frac{\sin\big[(m-n)kx\big]}{(m-n)k} - \frac{\sin\big[(m+n)kx\big]}{(m+n)k}\right]_0^{2\pi/k}$$

$$= \frac{\sin\big[2\pi(m-n)\big]}{2(m-n)k} - \frac{\sin\big[2\pi(m+n)\big]}{2(m+n)k}$$

$$= 0 \quad \text{if } m \text{ and } n \text{ are integers, and } m \neq n$$

The case of $m = n$ follows from putting $\cos\big[(m-n)kx\big] = 1$ in the first step; or, equivalently, by integrating $\sin^2(nkx)$ by using the double-angle formula.

$$\int_0^{2\pi/k} \sin^2(nkx)\,dx = \frac{\pi}{k}$$

$$\therefore \int_0^{2\pi/k} \sin(mkx)\sin(nkx)\,dx = \begin{cases} \dfrac{\pi}{k} & \text{if } m = n \\ 0 & \text{otherwise} \end{cases}$$

Similarly, $\quad \displaystyle\int_0^{2\pi/k} \cos(mkx)\cos(nkx)\,dx = \begin{cases} \dfrac{\pi}{k} & \text{if } m = n \\ 0 & \text{otherwise} \end{cases}$

$$\lambda = \frac{2\pi}{k}$$

*Fourier:* $\quad f(x) \approx \tfrac{1}{2}a_0 + a_1\cos(kx) + a_2\cos(2kx) + a_3\cos(3kx) + \cdots$

$$+ b_1\sin(kx) + b_2\sin(2kx) + b_3\sin(3kx) + \cdots$$

Multiplying both sides of the Fourier series by $\sin(nkx)$ and integrating over $0 \leqslant x \leqslant 2\pi/k$, and using the orthogonality relationships just derived, we obtain

$$\int_0^{2\pi/k} f(x)\sin(nkx)\,dx = 0 + b_n\int_0^{2\pi/k}\sin^2(nkx)\,dx + 0 = \frac{\pi}{k}b_n$$

$$\int_0^{2\pi/k} f(x)\,dx = \tfrac{1}{2}a_0\underbrace{\int_0^{2\pi/k}dx}_{\pi a_0/k} + 0$$

$$\therefore b_n = \frac{k}{\pi}\int_0^{2\pi/k} f(x)\sin(nkx)\,dx \qquad \text{Similarly,} \quad a_n = \frac{k}{\pi}\int_0^{2\pi/k} f(x)\cos(nkx)\,dx$$

One of the important ideas in this example is that of orthogonal functions. While orthogonality has a natural geometrical interpretation for vectors, namely that of perpendicular directions (having an angular separation of 90°), an equivalent physical picture for functions is not so obvious. Nevertheless, we can develop an analogy between vector manipulations and functional ones to gain a better appreciation of orthogonality in algebraic terms.

Two vectors are said to be orthogonal if their scalar product is nought

$$\mathbf{e}_i \cdot \mathbf{e}_j = \mathbf{e}_i^\mathsf{T} \mathbf{e}_j = 0 \quad \text{if } i \neq j$$

Likewise, two functions are called orthogonal if the integral of their product over some specified range, $\alpha \leqslant x \leqslant \beta$, is zero

$$\int_\alpha^\beta \mathbf{e}_i(x)\, \mathbf{e}_j(x)\, \mathrm{d}x = 0 \quad \text{if } i \neq j$$

$$\mathbf{a}^\mathsf{T} \mathbf{b} \;\rightarrow\; \mathbf{a}^{*\mathsf{T}} \mathbb{W} \, \mathbf{b}$$

This definition of a scalar product between functions can be generalized to include a 'weighting function', $w(x)$, as can the dot product between vectors with a 'metric' matrix $\mathbb{W}$; in the simplest case, $w(x) = 1$ and $\mathbb{W} = \mathbb{I}$. The first term in the scalar product can even be written as a complex conjugate, $\mathbf{e}_i^*$ or $\mathbf{e}_i(x)^*$, to allow for the possibility that the vectors or functions are not real.

Just as an $N$-dimensional vector, $\mathbf{X}$, can be decomposed into a linear combination of $N$ orthogonal basis vectors

$$\mathbf{X} = a_1\, \mathbf{e}_1 + a_2\, \mathbf{e}_2 + a_3\, \mathbf{e}_3 + \cdots + a_N\, \mathbf{e}_N$$

so too can a function, $f(x)$, be written as a linear combination of orthogonal basis functions

$$f(x) = a_1\, \mathbf{e}_1(x) + a_2\, \mathbf{e}_2(x) + a_3\, \mathbf{e}_3(x) + \cdots$$

$$\mathbf{X} \cdot \mathbf{e}_j = a_j\, \mathbf{e}_j \cdot \mathbf{e}_j$$
$$+ \underbrace{\sum_{i \neq j} a_i\, \mathbf{e}_i \cdot \mathbf{e}_j}_{0}$$

The $j^{\text{th}}$ coefficient, $a_j$, can be ascertained by taking the dot product of $\mathbf{X}$ with $\mathbf{e}_j$, or its functional analogue of multiplying $f(x)$ by $\mathbf{e}_j(x)$ and integrating between $\alpha \leqslant x \leqslant \beta$

$$a_j = \frac{\mathbf{X} \cdot \mathbf{e}_j}{|\mathbf{e}_j|^2} \qquad \text{or} \qquad a_j = \int_\alpha^\beta f(x)\, \mathbf{e}_j(x)\, \mathrm{d}x \;\Big/\; \int_\alpha^\beta \left[\mathbf{e}_j(x)\right]^2 \mathrm{d}x$$

If the basis vectors and functions are *orthonormal*, so that they are normalized as well as orthogonal, then these formulae can be simplified by omitting the denominators (because they are then equal to unity).

Given the discussion above, we can view a Fourier series as an example of a general procedure whereby $f(x)$ is expanded in terms of an orthogonal set of basis functions. Such a manipulation turns out to be very useful in science and engineering. Depending on the problem at hand, other commonly used orthogonal basis functions include *Bessel* functions, *Hermite* polynomials, *Laguerre* polynomials, *Legendre* polynomials, and spherical harmonics.

**(2)** For the Fourier expansion of eqn (15.1), derive Parseval's identity

$$\frac{1}{\pi}\int_{-\pi}^{\pi}\big[f(x)\big]^2\,dx \;=\; \tfrac{1}{2}a_0^2 + \sum_{n=1}^{\infty}\big(a_n^2 + b_n^2\big)$$

*Fourier* $(k=1)$:  $f(x) = \tfrac{1}{2}a_0 + a_1\cos x + a_2\cos 2x + a_3\cos 3x + \cdots$
$$+\ b_1\sin x + b_2\sin 2x + b_3\sin 3x + \cdots$$

Multiplying the left- and right-hand sides by themselves and integrating over the period $0 \leqslant x \leqslant 2\pi$, the orthogonality of the sines and cosines leads to

$$a_0 b_n \int_{0}^{2\pi}\sin(nx)\,dx = 0$$
$$\text{for } n=1,2,3,\ldots$$
$$\text{(and so on)}$$

$$\int_{0}^{2\pi}\big[f(x)\big]^2\,dx = \int_{0}^{2\pi}\Big[\tfrac{1}{4}a_0^2 + a_1^2\cos^2 x + b_1^2\sin^2 x + a_2^2\cos^2 2x + \cdots\Big]\,dx + 0$$

$$= \tfrac{\pi}{2}a_0^2 + \pi\big(a_1^2 + b_1^2 + a_2^2 + b_2^2 + a_3^2 + b_3^2 + \cdots\big)$$

$$\therefore\ \frac{1}{\pi}\int_{0}^{2\pi}\big[f(x)\big]^2\,dx = \frac{1}{\pi}\int_{-\pi}^{\pi}\big[f(x)\big]^2\,dx = \tfrac{1}{2}a_0^2 + \sum_{n=1}^{\infty}\big(a_n^2 + b_n^2\big)$$

The formula holds for an integral over $-\pi \leqslant x \leqslant \pi$ just as well as for $0 \leqslant x \leqslant 2\pi$ because the orthogonality for sines and cosines, and the implied periodicity of $f(x)$, relies on an integral over one repeat interval.

**(3)** A triangular wave is represented by

$$f(x) = \begin{cases} x & \text{for } 0 < x < \pi \\ -x & -\pi < x < 0 \end{cases} \qquad \text{and} \qquad f(x) = f(x + 2\pi)$$

Show that its Fourier series representation is given by

$$f(x) = \frac{\pi}{2} - \frac{4}{\pi}\sum_{n=0}^{\infty}\frac{\cos\big[(2n+1)x\big]}{(2n+1)^2}$$

Period $\lambda = 2\pi = \dfrac{2\pi}{k} \qquad \Rightarrow \qquad k = 1$

$\therefore$ *Fourier series*:  $f(x) = \tfrac{1}{2}a_0 + a_1\cos x + a_2\cos 2x + a_3\cos 3x + \cdots$
$$+\ b_1\sin x + b_2\sin 2x + b_3\sin 3x + \cdots$$

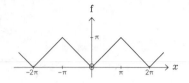

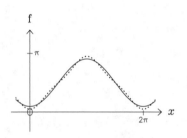

1$^{st}$– and 2$^{nd}$–order Fourier approximations

$$f(-x) = f(x) \quad \Rightarrow \quad b_n = \frac{1}{\pi} \int_{-\pi}^{\pi} f(x) \sin(nx)\, dx = 0$$

$$a_n = \frac{1}{\pi} \int_{-\pi}^{\pi} f(x) \cos(nx)\, dx = \frac{2}{\pi} \int_{0}^{\pi} x \cos(nx)\, dx$$

$$= \frac{2}{\pi} \left[ x\, \frac{\sin(nx)}{n} \right]_{0}^{\pi} - \frac{2}{n\pi} \int_{0}^{\pi} \sin(nx)\, dx$$

$$= 0 + \frac{2}{n\pi} \left[ \frac{\cos(nx)}{n} \right]_{0}^{\pi} = \frac{2\left[\cos(n\pi) - 1\right]}{n^2\, \pi}$$

$$\therefore \quad a_n = \begin{cases} \dfrac{-4}{n^2\, \pi} & \text{for } n = 1, 3, 5, \ldots \\[2mm] 0 & n = 2, 4, 6, \ldots \end{cases}$$

$$a_0 = \frac{2}{\pi} \int_{0}^{\pi} x\, dx = \frac{2}{\pi} \left[ \frac{x^2}{2} \right]_{0}^{\pi} = \pi$$

$$\therefore \quad f(x) = \frac{\pi}{2} - \frac{4}{\pi} \left( \frac{\cos x}{1^2} + \frac{\cos 3x}{3^2} + \frac{\cos 5x}{5^2} + \frac{\cos 7x}{7^2} + \cdots \right)$$

i.e. *Fourier series:* $\quad f(x) = \dfrac{\pi}{2} - \dfrac{4}{\pi} \displaystyle\sum_{n=0}^{\infty} \dfrac{\cos\left[(2n+1)x\right]}{(2n+1)^2}$

---

**(4)** For $\alpha > 0$, calculate the Fourier transforms of

$$\text{(a)} \ \ f(x) = \begin{cases} e^{-\alpha x} & \text{for } x \geqslant 0 \\ 0 & \text{otherwise} \end{cases}, \quad \text{(b)} \ \ g(x) = e^{-\alpha |x|}$$

---

(a) $\quad F(k) = \displaystyle\int_{-\infty}^{\infty} f(x)\, e^{-ikx}\, dx = \int_{0}^{\infty} e^{-(\alpha + ik)x}\, dx$

$$= \left[ -\frac{e^{-\alpha x}\, e^{-ikx}}{\alpha + ik} \right]_{0}^{\infty}$$

$$= 0 + \frac{1}{\alpha + ik} = \frac{\alpha - ik}{\alpha^2 + k^2}$$

We have omitted a factor of $2\pi$ in the definition of the Fourier transform here compared to eqn (15.6), so the corresponding inverse transform is

$$f(x) = \frac{1}{2\pi} \int_{-\infty}^{\infty} F(k)\, e^{ikx}\, dk$$

rather than eqn (15.5).

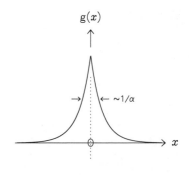

$g(x)$

$\leftarrow \sim 1/\alpha$

(b)  $G(k) = \int_{-\infty}^{\infty} g(x)\, e^{-ikx}\, dx = \int_{-\infty}^{0} e^{\alpha x}\, e^{-ikx}\, dx + \underbrace{\int_{0}^{\infty} e^{-\alpha x}\, e^{-ikx}\, dx}_{F(k)}$

Putting $x = -t$,  $\int_{-\infty}^{0} e^{\alpha x}\, e^{-ikx}\, dx = \int_{0}^{\infty} e^{-\alpha t}\, e^{ikt}\, dt = \left[F(k)\right]^{*}$

$\therefore \quad G(k) = F(k) + \left[F(k)\right]^{*} = 2\,\mathcal{R}e\,\{F(k)\} = \dfrac{2\alpha}{\alpha^2 + k^2}$

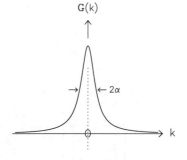

$G(k)$

$\leftarrow 2\alpha$

$k$

While $f(x)$ is a real function, the closely-related $g(x)$ is also symmetric. Hence, as noted in section (15.5), $G(k)$ is real and symmetric whereas $F(k)$ is merely conjugate symmetric. The explicit calculation of the Fourier integral to obtain $g(x)$ from $G(k)$, however, requires a knowledge of analysis techniques in the complex plane that are beyond the scope of this primer.

The reciprocal feature of Fourier transforms mentioned in section (15.4) can also be seen here: the width of the peak in $g(x)$ is inversely related to the spread of the Lorentzian function $G(k)$. This is even clearer with the Fourier transform of the bell-shaped Gaussian function

$$\int_{-\infty}^{\infty} \exp\left(-\frac{x^2}{2\sigma^2}\right) e^{-ikx}\, dx \propto \exp\left(-\frac{\sigma^2 k^2}{2}\right)$$

because both the function and its Fourier transforms have exactly the same shape but reciprocal widths (a *standard deviation* of $\sigma$ in $x$ versus $1/\sigma$ in k).

---

**(5)** With reference to eqns (15.13) and (15.14), calculate the diffraction pattern, $I(q)$, from a pair of slits of width $w$ that are separated by a distance $d$ (which is much greater than $w$):

$$A(x) = \begin{cases} 1 & \text{for } \left|x \pm \tfrac{1}{2}d\right| \leqslant \tfrac{1}{2}w \\ 0 & \text{otherwise} \end{cases}$$

$$\frac{\psi(q)}{\psi_0} = \int_{-\infty}^{\infty} A(x)\,e^{iqx}\,dx = \int_{-\frac{d}{2}-\frac{w}{2}}^{-\frac{d}{2}+\frac{w}{2}} e^{iqx}\,dx + \int_{\frac{d}{2}-\frac{w}{2}}^{\frac{d}{2}+\frac{w}{2}} e^{iqx}\,dx$$

$$= \left[\frac{e^{iqx}}{iq}\right]_{-\frac{d}{2}-\frac{w}{2}}^{-\frac{d}{2}+\frac{w}{2}} + \left[\frac{e^{iqx}}{iq}\right]_{\frac{d}{2}-\frac{w}{2}}^{\frac{d}{2}+\frac{w}{2}}$$

$$= \frac{e^{-iqd/2}\left(e^{iqw/2} - e^{-iqw/2}\right)}{iq} + \frac{e^{iqd/2}\left(e^{iqw/2} - e^{-iqw/2}\right)}{iq}$$

$$= \frac{2\left(e^{iqd/2} + e^{-iqd/2}\right)\sin(qw/2)}{q} = 2\,w\,\cos\left(\frac{qd}{2}\right)\mathrm{sinc}\left(\frac{qw}{2}\right)$$

where $\mathrm{sinc}\,\theta = \sin\theta/\theta$. Hence, the diffraction pattern is

$$I(q) = |\psi(q)|^2 = 4\,w^2 I_0 \cos^2\left(\frac{qd}{2}\right)\mathrm{sinc}^2\left(\frac{qw}{2}\right)$$

where $I_0 = |\psi_0|^2$. As noted in section (15.6), this result could also be obtained with the convolution theorem from a knowledge of the diffraction patterns of a Young's double slit and a single wide slit.

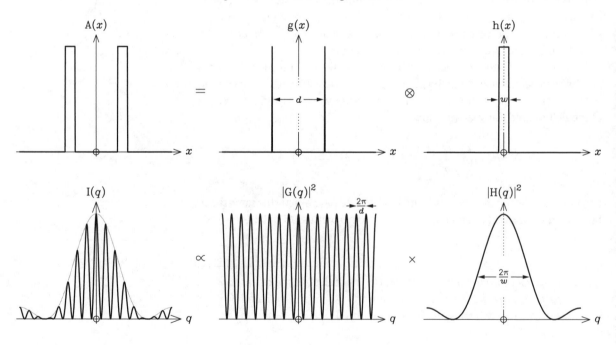

# 16 Data analysis

## 16.1 Introduction

One of the things that sets science apart from other systems of trying to make sense of the world around us is the use of empirical evidence: we conduct experiments to yield data that guide us towards useful models for understanding physical phenomena. A scientific theory must therefore be able to make predictions that can be compared with observations.

Since the amount of data obtained in an experiment is limited, and subject to noise in the measurement process and uncertainty in the parameters required for their analysis, the conclusions reached have to be couched in conditional terms with an appropriate statement about their reliability. How this is done is the focus of this chapter, but the topic is both important and large. Here we will only be able to scratch the surface, by considering a few simple but common examples, and refer the reader to a book dedicated to data analysis that is written in a style very similar to this primer.[†]

Before beginning the quantitative discussion, we should point out that a qualitative visual assessment of the experimental measurements may be sufficient to address the issue at hand. In the analysis of the chemical rate equations in question (2) of chapter 13, for example, we found that the variation in the concentration of a reactant, $c$, with time, $t$, was

$$c = c_0 e^{-kt} \qquad \Longleftrightarrow \qquad \ln c = \ln c_0 - kt$$

for first-order reactions, with $c_0$ being the initial concentration and k the rate constant, whereas

$$c = \frac{c_0}{1 + c_0 kt} \qquad \Longleftrightarrow \qquad \frac{1}{c} = \frac{1}{c_0} + kt$$

for second-order ones. Given measurements of $c$ through the course of the reaction, the simplest way of differentiating between the two cases is to plot $\ln c$ and $1/c$ versus time: the logarithmic graph will be a straight line if the reaction is first-order and the reciprocal will be linear if its second-order; in both cases, the slope gives the rate constant. Even if such a qualitative assessment of the data provides a convincing answer for the order of the chemical reaction, what can be said quantitatively about the value of k? How should we obtain a best estimate and a measure of its reliability?

[†] **Data Analysis: a Bayesian tutorial**
D. S. Sivia with J. Skilling (2006)
Oxford University Press

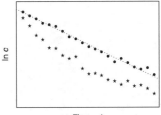

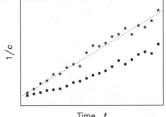

## 16.2 Inference: the basic rules

Drawing inferences from incomplete and uncertain information, as a basis for making decisions, is a task that we face all the time in our daily lives: should I take an umbrella, is it safe to cross the road, is this just a passing illness or should I visit the doctor, and so on. In the modern scientific context, philosophers started to think about how to reason in situations where we cannot argue with certainty about 300 years ago. Laplace laid the foundations of a coherent mathematical theory for addressing such questions in his 1812 treatise, *Théorie analytique des probabilités*. As he put it, "It is remarkable that a science, which commenced with a consideration of games of chance, should be elevated to the rank of the most important subjects of human knowledge."

Despite Laplace's success in tackling problems in celestial mechanics, medicine, and even jurisprudence, his development of probability theory was rejected by many soon after his death. The idea of a probability representing how much we believe that something is true, based on the evidence at hand, seemed too vague and subjective to be the basis of a rigorous theory. So probability was re-defined as the long-run relative frequency with which an event occurred, given (infinitely) many repeated (experimental) trials, and seen as an objective tool for dealing with random phenomena.

Unfortunately the strict frequency interpretation does not allow probability theory to be used directly to address inference problems, and this led to the de-velopment of 'conventional' statistics to deal with the shortcoming. Although the frequency viewpoint has been dominant since the mid-19th century, there has been a resurgence in the older 'Bayesian' way of thinking over the past few decades. Indeed, it underlies much of the work on artificial intelligence (AI) and machine learning.

*"Probability theory is nothing but common sense reduced to calculation."*
— Laplace, 1814

The rules of probability theory are the same for the frequency definition and the degree-of-belief viewpoint. The difference is one of interpretation, but this affects which problems are deemed directly amenable to a probabilistic analysis. Since Laplace thought that "Life's most important questions are, for the most part, nothing but probability problems", he considered probability theory to be applicable in all situations involving uncertainty. With the assignment of zero or one for the probability of some proposition, '$X$', being true defining certainty,

$$0 \leqslant \text{prob}(X|I) \leqslant 1 \tag{16.1}$$

where the vertical bar means 'given', so that all items to the right of this condition-ing symbol are taken as being true, and $I$ denotes the relevant background infor-mation at hand. As a probability represents a state-of-knowledge in this frame-work, all probabilities are inherently conditional. The probability assigned to the proposition 'it will rain this afternoon', for example, will depend on whether there are dark clouds or a clear blue sky in the morning; it will also be affected by access to the current air pressure trend and recent satellite imagery.

Apart from eqn (16.1), there are only two basic rules in probability theory:

$$\text{prob}(X|I) + \text{prob}(\overline{X}|I) = 1 \tag{16.2}$$

where $\overline{X}$ stands for the converse of $X$, that $X$ is false, and

$$\text{prob}(X, Y \mid I) = \text{prob}(X \mid Y, I) \times \text{prob}(Y \mid I) \tag{16.3}$$

where $Y$ asserts the truth of a second proposition and the comma is read as the conjunction 'and' (like the intersection symbol $\cap$ in set theory).

Many other results can be derived from these *sum* and *product* rules. Among the most useful corollaries are *Bayes' theorem*

$$\text{prob}(X \mid Y, I) = \frac{\text{prob}(Y \mid X, I) \times \text{prob}(X \mid I)}{\text{prob}(Y \mid I)} \tag{16.4}$$

which follows from eqn (16.3) and the equivalent expression for $\text{prob}(Y, X \mid I)$, as they are different decompositions of the same probability (of $X$ and $Y$ both being true), and *marginalization*

$$\text{prob}(Y \mid I) = \text{prob}(Y, X \mid I) + \text{prob}(Y, \overline{X} \mid I) \tag{16.5}$$

If the alternatives are not simply $X$ and $\overline{X}$, but a set of N *mutually exclusive* and *exhaustive* possibilities (if one is true the others must be false, but one of them has to be true)

$$X_j \equiv \text{proposition } X_j \text{ is true}, \quad \text{for } j = 1, 2, 3, \ldots, N$$

then eqn (16.2) becomes

$$\sum_{j=1}^{N} \text{prob}(X_j \mid I) = 1 \tag{16.6}$$

and $\text{prob}(X_j \mid I)$, for $j = 1$ to N, is called a *probability distribution function* for the $X_j$. Similarly, eqn (16.5) generalizes to

$$\text{prob}(Y \mid I) = \sum_{j=1}^{N} \text{prob}(Y, X_j \mid I) \tag{16.7}$$

$$\text{prob}(X_j \mid Y, I) = \frac{\text{prob}(Y \mid X_j, I) \times \text{prob}(X_j \mid I)}{\text{prob}(Y \mid I)}$$

For the earlier elementary case, $X_1 = X$, $X_2 = \overline{X}$, and N $= 2$.

Now suppose that $x$ is a quantity that can take any value between the bounds of $x_{\min}$ and $x_{\max}$, such as a person's height, and define

$$X_j \equiv (j-1)\,\delta x \leqslant x - x_{\min} < j\,\delta x, \quad \text{for } j = 1, 2, 3, \ldots, N$$

where

$$\delta x = \frac{x_{\max} - x_{\min}}{N}$$

Then, these $X_j$ constitute a set of mutually exclusive and exhaustive alternatives. In the limit $N \to \infty$, $\text{prob}(X_j \mid I)$ becomes a continuous function of $x$ called a *probability density function*:

$$\text{pdf}(X = x \mid I) = \lim_{\delta x \to 0} \frac{\text{prob}(x \leqslant X < x + \delta x \mid I)}{\delta x}$$

and the probability that $x$ lies in a finite range between $x_1$ and $x_2$ is given by

$$\text{prob}(x_1 \leqslant X < x_2 \,|\, I) = \int_{x_1}^{x_2} \text{pdf}(X\,|\,I)\, \mathrm{d}X \qquad (16.8)$$

The normalization condition then takes the form

$$\int_{x_{\min}}^{x_{\max}} \text{pdf}(X\,|\,I)\, \mathrm{d}X = 1 \qquad (16.9)$$

with $\text{pdf}(X\,|\,I) = 0$ for $x < x_{\min}$ and $x > x_{\max}$. Bayes' theorem becomes

$$\text{pdf}(X\,|\,Y,I) = \frac{\text{prob}(Y\,|\,X,I) \times \text{pdf}(X\,|\,I)}{\text{prob}(Y\,|\,I)} \qquad (16.10)$$

and the denominator is given by the marginal integral

$$\text{prob}(Y\,|\,I) = \int_{x_{\min}}^{x_{\max}} \text{pdf}(Y, X\,|\,I)\, \mathrm{d}X \qquad (16.11)$$

## 16.3 Common probability distributions

Equations (16.2) and (16.3), and their corollaries, define how probabilities are related to each other but they do not specify how any of them should be assigned. This is still an open question in general, as the state of knowledge (or ignorance) that we are trying to represent can take a myriad of different forms (often rather vague). There are some insightful ideas based on arguments of consistency, that if two people have the same information they should assign the same probability, but here we simply list some cases that are met frequently.

### Uniform distribution

The simplest assignment is that of a uniform distribution

$$\text{prob}(X_j\,|\,I) = \frac{1}{N} \quad \text{for } j = 1, 2, 3, \ldots, N \qquad (16.12)$$

or, in the continuous case,

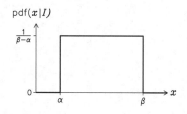

$$\text{pdf}(X = x\,|\,I) = \begin{cases} \dfrac{1}{x_{\max} - x_{\min}} & \text{for } x_{\min} \leqslant x \leqslant x_{\max} \\[2mm] 0 & \text{otherwise} \end{cases} \qquad (16.13)$$

The uniform distribution encodes the information that we can enumerate an exhaustive set of mutually exclusive possibilities, or know the range in which the value of a parameter might lie, but have little else to go on.

## Normal distribution

The quadratic exponent of the Gaussian function

$$\text{pdf}(X = x \mid \mu, \sigma, I) = \frac{1}{\sigma\sqrt{2\pi}} \exp\left[-\frac{(x-\mu)^2}{2\sigma^2}\right] \tag{16.14}$$

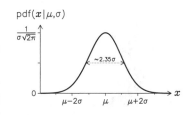

gives it the classical bell-shaped curve that is encountered most often in probabilistic discussion. Hence, its status as the 'normal' distribution. It is symmetric with respect to the maximum at $x = \mu$, called the *mode*, and has a full width at half maximum of about $2.35\sigma$.

The *mean*, average or *expectation* value of $x$, variously denoted by $<x>$, $\bar{x}$ or $\mathbb{E}(X)$, is equal to $\mu$

$$\mathbb{E}(X) = \int_{-\infty}^{+\infty} X \, \text{pdf}(X \mid \mu, \sigma, I) \, dX = \mu \tag{16.15}$$

The *variance* of $x$, which is the expected value of the square of the deviation of $x$ from its mean, is equal to $\sigma^2$

$$\text{var}(X) = \mathbb{E}\big[(X-\mu)^2\big] \tag{16.16}$$

$$= \int_{-\infty}^{+\infty} (X-\mu)^2 \, \text{pdf}(X \mid \mu, \sigma, I) \, dX = \sigma^2$$

Incidentally, the expansion of the quadratic in the integrand, and a simplification of the three resultant integrals, shows that the variance can be expressed as the difference of two expectations

$$\text{var}(X) = \mathbb{E}(X^2) - \big[\mathbb{E}(X)\big]^2 \tag{16.17}$$

The square root of the variance, $\sigma$, is known as the *standard deviation*. The probability that $|x - \mu| \leqslant \sigma$ is about 68%, and over 95% for $2\sigma$. With this in mind, eqn (16.14) can be summarized by saying that $x = \mu \pm \sigma$.

The normal distribution can be viewed as the one that best represents our state of knowledge about the value of $x$ when only given information about its mean and variance. It can also be derived as a limiting form in other cases.

## Binomial distribution

A consideration of $n$ independent trials, conducted under identical conditions, with just two possible outcomes, generically called 'success' and 'failure', leads to the binomial distribution for the probability of obtaining $r$ successes

$$\text{prob}(r \mid n, p, I) = {}^nC_r \, p^r \, (1-p)^{n-r} \tag{16.18}$$

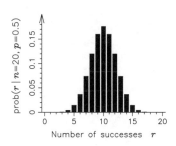

for $r = 0, 1, 2, \ldots, n$. Here $0 \leqslant p \leqslant 1$ is the probability of success in each trial, so $1-p$ is the chance of failure, and the binomial coefficient gives the number of ways of choosing $r$ objects from $n$ when the order does not matter.

$${}^nC_r = \frac{n!}{r! \, (n-r)!}$$

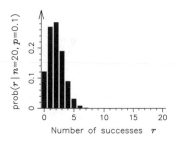

The expected number of successes is given by

$$\mathbb{E}(r) = \sum_{r=0}^{n} r \, \text{prob}(r|n,p,I) = np \qquad (16.19)$$

with a mean square deviation of

$$\text{var}(r) = \sum_{r=0}^{n} r^2 \, \text{prob}(r|n,p,I) - \left[\mathbb{E}(r)\right]^2 = np(1-p) \qquad (16.20)$$

As $n$ increases, eqn (16.18) starts to resemble a discrete version of the normal distribution: $r \approx np \pm \sqrt{np(1-p)}$ when $n$ is much greater than the larger of $p/(1-p)$ and its reciprocal.

### Poisson distribution

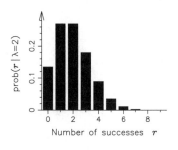

If the probability of success in each trial is very small, $p \to 0$, but the number of attempts is huge, $n \to \infty$, while $np = \lambda$ remains finite, then eqn (16.18) takes the form of a Poisson distribution

$$\text{prob}(r|\lambda,I) = \frac{\lambda^r \, e^{-\lambda}}{r!} \qquad (16.21)$$

for $r = 0, 1, 2, 3, \ldots$. Unsurprisingly, it can formally be shown to have a mean and variance equal to $\lambda$

$$\mathbb{E}(r) = \lambda \qquad \text{and} \qquad \text{var}(r) = \mathbb{E}\left[(r-\lambda)^2\right] = \lambda \qquad (16.22)$$

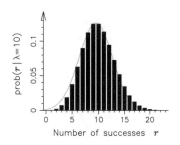

If $\lambda \gg 1$, eqn (16.21) can be approximated reasonably by a normal distribution: $r \approx \lambda - \frac{1}{2} \pm \sqrt{\lambda}$.

The binomial and Poisson distributions can also be regarded as assignments that best represent some limited knowledge about $r$. The expected number of successes in $n$ trials, $\mathbb{E}(r|n) = \lambda$, for the binomial

$$\text{prob}(r|n,\lambda,I) = {}^nC_r \left(\frac{\lambda}{n}\right)^r \left(1-\frac{\lambda}{n}\right)^{n-r}$$

where $r = 0, 1, 2, \ldots, n$; and $\mathbb{E}(r) = \lambda$ over a continuous interval (spatial or temporal) for eqn (16.21).

## 16.4 Parameter estimation

The estimation of a quantity, $\mu$ say, from repeated measurements of it is one of the most elementary problems in data analysis. If there are N data $\{d_k\}$, with $k$ going from 1 to N, each subject to a normal uncertainty of $\sigma$, then the inference about $\mu$ is described by $\text{pdf}(\mu|\{d_k\}, \sigma, I)$: a narrow distribution signals a high degree of confidence in the estimated value of $\mu$, whereas a broad one warns that it is poorly determined.

The desired probability, $\text{pdf}(\mu|\{d_k\},\sigma,I)$, can be related to others that are assigned more easily through Bayes' theorem:

$$\text{pdf}(\mu|\{d_k\},\sigma,I) \propto \text{prob}(\{d_k\}|\mu,\sigma,I) \times \text{pdf}(\mu|I) \qquad (16.23)$$

where the equality has been replaced by a proportionality due to the omission of the denominator, since it does not explicitly involve $\mu$, and the conditioning of $\text{pdf}(\mu|I)$ on $\sigma$ has been dropped, as it tells us little about $\mu$ without the data. The specification of $\sigma$ as a normal uncertainty implies that

$$\text{prob}(d_k|\mu,\sigma,I) = \frac{1}{\sigma\sqrt{2\pi}} \exp\left[-\frac{(d_k-\mu)^2}{2\sigma^2}\right] \qquad (16.24)$$

while an assumption that the measurements are *independent*, so that the error in one does not influence the fluctuation of another, means that

$$\text{prob}(\{d_k\}|\mu,\sigma,I) = \prod_{k=1}^{N} \text{prob}(d_k|\mu,\sigma,I) \qquad (16.25)$$

The latter follows from the repeated use of the independence relationship

$$\prod_{k=1}^{N} A_k = A_1 A_2 A_3 \dots A_N$$

$$\text{prob}(X,Y|I) = \text{prob}(X|I) \times \text{prob}(Y|I)$$

the form to which eqn (16.3) reduces when $\text{prob}(X|Y,I)=\text{prob}(X|I)$, indicating that a knowledge of $Y$ is not informative about $X$. With the assignment of a uniform distribution for $\text{pdf}(\mu|I)$ over a suitably large (but finite) range, to encode gross initial ignorance, eqns (16.23)–(16.25) yield

$$L = \ln\left[\text{pdf}(\mu|\{d_k\},\sigma,I)\right] = C - \frac{1}{2\sigma^2}\sum_{k=1}^{N}(d_k-\mu)^2 \qquad (16.26)$$

where all the omitted proportionality constants have been combined and give an additive one, $C$, on taking the logarithm.

The expansion of $L$ in a Taylor series about its maximum at $\mu=\mu_0$

$$L = L(\mu_0) + \frac{1}{2}\frac{d^2L}{d\mu^2}\bigg|_{\mu_0}(\mu-\mu_0)^2, \quad \text{where} \quad \frac{dL}{d\mu}\bigg|_{\mu_0} = 0$$

$$\frac{dL}{d\mu} = \frac{1}{\sigma^2}\sum_{k=1}^{N}(d_k-\mu)$$

shows that $\text{pdf}(\mu|\{d_k\},\sigma,I)=e^L$ is a normal distribution, as all the terms in $L$ beyond the quadratic are zero. A comparison with the logarithm of eqn (16.14) leads us to summarize the inference by

$$\mu = \mu_0 \pm \left[-\frac{d^2L}{d\mu^2}\bigg|_{\mu_0}\right]^{-\frac{1}{2}} = \frac{1}{N}\sum_{k=1}^{N}d_k \pm \frac{\sigma}{\sqrt{N}} \qquad (16.27)$$

$$\frac{d^2L}{d\mu^2} = -\frac{1}{\sigma^2}\underbrace{\sum_{k=1}^{N}1}_{N}$$

so that the best estimate of $\mu$ is given by the arithmetic mean of the data and the confidence in this value improves with the square root of their number.

If the magnitude of the noise in the measurements, $\sigma$, is not known, then the inference about $\mu$ is described by the distribution $\text{pdf}(\mu|\{d_k\},I)$. As earlier, the probability sought can be related to others that are assigned more easily through the basic rules discussed in section (16.2):

$$\text{pdf}(\mu|\{d_k\}, I) \ \propto \ \text{pdf}(\mu|I) \times \text{prob}(\{d_k\}|\mu, I)$$

$$\propto \ \text{pdf}(\mu|I) \underbrace{\int_0^\infty \text{prob}(\{d_k\}|\mu, \sigma, I)\, \text{pdf}(\sigma|I)\, \text{d}\sigma}_{\text{prob}(\{d_k\}, \sigma|\mu, I)}$$

where the conditioning of $\text{pdf}(\sigma|I)$ on $\mu$ has been dropped, as it tells us little about $\sigma$ without the data, and the lower limit of the marginal integral has been set to zero because $\sigma$ cannot be negative. Using eqns (16.24) and (16.25), and making uniform assignments for $\text{pdf}(\mu|I)$ and $\text{pdf}(\sigma|I)$ to encode gross initial ignorance, the integration over $\sigma$ yields

$$L = \ln\left[\text{pdf}(\mu|\{d_k\}, I)\right] = \mathcal{C} - \frac{(N-1)}{2}\ln\left[\sum_{k=1}^N (d_k - \mu)^2\right] \qquad (16.28)$$

Since the derivatives higher than $\text{d}^2 L/\text{d}\mu^2$ are no longer zero, $\text{pdf}(\mu|\{d_k\}, I)$ is not a normal distribution. Nevertheless, it can be approximated reasonably well by one when $N \gg 1$. The exponential of the quadratic Taylor series of $L$ about its maximum gives

$$\mu \approx \underbrace{\frac{1}{N}\sum_{k=1}^N d_k}_{\mu_0} \pm \frac{S}{\sqrt{N}}, \quad \text{where } \ S^2 = \frac{1}{N-1}\sum_{k=1}^N (d_k - \mu_0)^2 \qquad (16.29)$$

The best estimate of $\mu$ is still given by the arithmetic mean of the data, therefore, but the unknown noise level is replaced by an estimate obtained from the measurements.

### Multivariate generalization

The formalism used above can be generalized to deal with the case of many variables by adopting matrix-vector notation. For a model with M parameters, for example, let them be the components of an M-dimensional vector $\mathbf{x}$. Similarly, denote the data by an N-dimensional vector $\mathbf{d}$. The inference is then defined by $\text{pdf}(\mathbf{x}|\mathbf{d}, I)$, called the *posterior* probability, as it represents our state of knowledge about $\mathbf{x}$ in the light of the data (and the relevant background information, and analysis assumptions, subsumed in $I$). This is related to the *prior* probability, $\text{pdf}(\mathbf{x}|I)$, and the *likelihood* function, $\text{prob}(\mathbf{d}|\mathbf{x}, I)$, through Bayes' theorem

$$\mathbf{x} = \begin{pmatrix} x_1 \\ x_2 \\ \vdots \\ x_M \end{pmatrix}$$

$$\text{pdf}(\mathbf{x}|\mathbf{d}, I) = \frac{\text{prob}(\mathbf{d}|\mathbf{x}, I) \times \text{pdf}(\mathbf{x}|I)}{\text{prob}(\mathbf{d}|I)} \qquad (16.30)$$

where the normalizing denominator

$$\mathbf{d} = \begin{pmatrix} d_1 \\ d_2 \\ \vdots \\ d_N \end{pmatrix}$$

$$\text{prob}(\mathbf{d}|I) = \int\!\!\int \cdots \int \text{prob}(\mathbf{d}|\mathbf{x}, I)\, \text{pdf}(\mathbf{x}|I)\, \text{d}^M \mathbf{x} \qquad (16.31)$$

is often omitted in parameter estimation problems, and the equality in eqn (16.30) replaced with a proportionality.

If the posterior probability has a dominant peak, spread symmetrically about a maximum at $\mathbf{x} = \mathbf{x}_0$, then it may be approximated reasonably well by a multivariate normal distribution

$$\text{pdf}(\mathbf{x}|\mathbf{d}, I) \propto \exp\left[-\tfrac{1}{2}(\mathbf{x} - \mathbf{x}_0)^{\mathsf{T}}\, \mathbb{H}\,(\mathbf{x} - \mathbf{x}_0)\right] \qquad (16.32)$$

where, following section (10.4), $\mathbf{x}_0$ and the *Hessian* matrix $\mathbb{H}$ are given by

$$\nabla L(\mathbf{x}_0) = 0 \quad \text{and} \quad \mathbb{H} = -\nabla\nabla L(\mathbf{x}_0) \qquad (16.33)$$

and $L = \ln\left[\text{pdf}(\mathbf{x}|\mathbf{d}, I)\right]$. Since $\mathbb{H}$ is the generalization of the $1/\sigma^2$ term in eqn (16.14), $\mathbb{H}^{-1} = \boldsymbol{\sigma}^2$ is called the *covariance* matrix. It can formally be shown that

$$\left(\boldsymbol{\sigma}^2\right)_{ij} = \left(\mathbb{H}^{-1}\right)_{ij} = \mathbb{E}\left[(x_i - x_{0i})(x_j - x_{0j})\right] \qquad (16.34)$$

for $i$ and $j = 1, 2, \ldots, M$. The diagonal terms of $\boldsymbol{\sigma}^2$ give the normal (marginal) uncertainties for the inferred parameters

$$x_i \approx (\mathbf{x}_0)_i \pm \sqrt{\left(\boldsymbol{\sigma}^2\right)_{ii}} \qquad (16.35)$$

and the off-diagonal terms specify the correlations between them. Covariances ($i \neq j$) are judged relative to the geometric mean of their associated variances, from the *correlation coefficient*

$$c_{ij} = \frac{\left(\boldsymbol{\sigma}^2\right)_{ij}}{\sqrt{\left(\boldsymbol{\sigma}^2\right)_{ii}\left(\boldsymbol{\sigma}^2\right)_{jj}}} = \frac{\mathbb{E}\left[(x_i - x_{0i})(x_j - x_{0j})\right]}{\sqrt{\mathbb{E}\left[(x_i - x_{0i})^2\right]\mathbb{E}\left[(x_j - x_{0j})^2\right]}}$$

which will lie between $-1$ and $+1$. When it is highly negative, one parameter tends to be underestimated when the other is overestimated; when significantly positive, both tend to be simultaneously high or low. A value close to zero represents the uncorrelated case.

## Simplifying approximations

Some of the most commonly used statistical prescriptions for parameter estimation follow from Bayes' theorem once simplifying approximations are made with regard to the prior and the likelihood function. A uniform prior makes the posterior probability proportional to the likelihood function, for example, so that the model parameters which make the given data most likely are also those that are most probable in the light of the measurements. This provides a justification for the method of *maximum likelihood*.

Assuming further that the data are subject to independent, additive, Gaussian noise, as in eqns (16.24) and (16.25), the posterior probability takes the form

$$L = \ln\left[\text{pdf}(\mathbf{x}|\mathbf{d}, I)\right] = C - \frac{1}{2}\sum_{k=1}^{N}\left(\frac{d_k - f_k}{\sigma_k}\right)^2 \qquad (16.36)$$

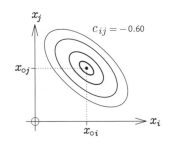

$c_{ij} = -0.60$

$$\left.\frac{\partial L}{\partial x_i}\right|_{\mathbf{x}_0} = 0$$

$$\left(\mathbb{H}\right)_{ij} = -\left.\frac{\partial^2 L}{\partial x_i\,\partial x_j}\right|_{\mathbf{x}_0}$$

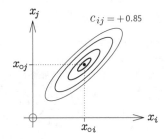

$c_{ij} = +0.85$

$$\mathbf{x} = \begin{pmatrix} m \\ c \end{pmatrix}$$

$$f_k = m\,x_k + c$$

where $\sigma_k$ is the uncertainty associated with the datum $d_k$ and $f_k$ is the corresponding predication for the measurement: $f_k = f(\mathbf{x}, k\,|\,I)$. The posterior probability is then maximized when the quadratic sum of the normalized *residuals*, $(d_k - f_k)/\sigma_k$, is as small as possible. An example of such a *least-squares* estimate, with $\sigma_k = \sigma$ (a constant), for the gradient and intercept of a straight line, can be found in question (8) of chapter 10.

## 16.5 Propagation of errors

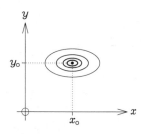

Given that $x = x_0 \pm \sigma_x$ and $y = y_0 \pm \sigma_y$, what can be said about the difference $x - y$, or the ratio $x/y$, or the logarithm $\ln x$, and so on? This type of question is addressed formally by using the rules of section (16.2) to make an appropriate change of variables: ascertain $\mathrm{pdf}(z\,|\,I)$ given $\mathrm{pdf}(x, y\,|\,I)$ where $z = f(x, y)$. In practice, the 'propagation of errors' is usually carried out with a linear Taylor series expansion of $z$ about $(x_0, y_0)$

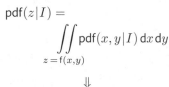

$$z = \underbrace{f(x_0, y_0)}_{z_0} + (x - x_0)\left.\frac{\partial f}{\partial x}\right|_{x_0, y_0} + (y - y_0)\left.\frac{\partial f}{\partial y}\right|_{x_0, y_0} + \cdots$$

under the condition that $\sigma_x$ and $\sigma_y$ are small enough for the higher-order terms to be neglected in the regions of high probability density. Moving $z_0 = f(x_0, y_0)$ over to the left and then squaring both sides

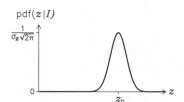

$$(\delta z)^2 \approx (\delta x)^2 \left.\frac{\partial f}{\partial x}\right|^2_{x_0, y_0} + (\delta y)^2 \left.\frac{\partial f}{\partial y}\right|^2_{x_0, y_0} + 2\,\delta x\,\delta y \left.\frac{\partial f}{\partial x}\right|_{x_0, y_0}\left.\frac{\partial f}{\partial y}\right|_{x_0, y_0}$$

where $\delta z = z - z_0$ and so on. Taking expectation values all the way through, and making the identification

$$\mathbb{E}\big[(\delta x)^2\big] = \sigma_x^2, \quad \mathbb{E}\big[(\delta y)^2\big] = \sigma_y^2 \quad \text{and} \quad \mathbb{E}\big[(\delta z)^2\big] = \sigma_z^2$$

we find that

$$\sigma_z^2 \approx \sigma_x^2 \left.\frac{\partial f}{\partial x}\right|^2_{x_0, y_0} + \sigma_y^2 \left.\frac{\partial f}{\partial y}\right|^2_{x_0, y_0} + 2\,\underbrace{\mathbb{E}\big[\delta x\,\delta y\big]}_{0} \left.\frac{\partial f}{\partial x}\right|_{x_0, y_0}\left.\frac{\partial f}{\partial y}\right|_{x_0, y_0} \tag{16.37}$$

In the absence of any cogent information on the correlation between $x$ and $y$, $\mathbb{E}\big[\delta x\,\delta y\big]$ is usually set to zero (the independence assumption). Thus, we have $z \approx z_0 \pm \sigma_z$.

For the sum and difference $z = x \pm y$, the error propagation formula of eqn (16.37) is exact because all the neglected terms are identically equal to zero. With $\partial f/\partial x = 1$, $\partial f/\partial y = \pm 1$, and $\mathbb{E}\big[\delta x\,\delta y\big] = 0$,

$$z_0 = x_0 \pm y_0 \quad \text{and} \quad \sigma_z = \sqrt{\sigma_x^2 + \sigma_y^2} \tag{16.38}$$

For the ratio $z = x/y$, the corresponding expression is

$$z_0 = \frac{x_0}{y_0} \quad \text{and} \quad \sigma_z = |z_0| \sqrt{\left(\frac{\sigma_x}{x_0}\right)^2 + \left(\frac{\sigma_y}{y_0}\right)^2} \tag{16.39}$$

which is valid for $\sigma_x \ll |x_0|$ and $\sigma_y \ll |y_0|$.

## 16.6 Model comparison

Hitherto we have assumed that there is just one theory under consideration, and have discussed issues related to the estimation of the associated model parameters. How should we deal with the case of competing hypotheses?

An example was met in section (16.1), where first- and second-order reactions were distinguished qualitatively by plotting the logarithm and reciprocal of the concentration data as a function of time to see which one more closely resembled a straight line. A choice between competing hypotheses, $H_1$ and $H_2$, can be made quantitatively by calculating the ratio of their posterior probabilities:

$$\frac{\text{prob}(H_1|\mathbf{d},I)}{\text{prob}(H_2|\mathbf{d},I)} = \frac{\text{prob}(H_1|I)}{\text{prob}(H_2|I)} \times \frac{\text{prob}(\mathbf{d}|H_1,I)}{\text{prob}(\mathbf{d}|H_2,I)} \tag{16.40}$$

where the right-hand side follows from applying Bayes' theorem to the numerator and denominator on the left. With the ratio of prior probabilities usually set to unity, the odds in favour of the competing hypotheses are determined by the $\text{prob}(\mathbf{d}|H,I)$ terms. This is the denominator factor in Bayes' theorem that tends to be omitted in parameter estimation problems, when a specific model is being considered. Equation (16.31) shows this to be the mean value of the likelihood function for the associated parameters, averaged over their prior probability, and is variously referred to as the 'global likelihood', 'prior predictive', or 'evidence'. It increases greatly if there is a significant improvement in the agreement with the data, but diminishes when the gain with a growing number of variables is meagre. If $\text{prob}(H_1|\mathbf{d},I)/\text{prob}(H_2|\mathbf{d},I) \gg 1$, then $H_1$ is preferred; $H_2$ is favoured when the posterior ratio is much smaller than 1; a value of order unity means that an informed judgment cannot be made.

## 16.7 Concluding remarks

Just like quantum mechanics, only a handful of simple data analysis problems can be solved analytically. In general, numerical methods need to be used. Very few people now write their own computer programs, as there are plenty of analysis software packages available. It should be borne in mind, however, that practical algorithms entail a compromise between speed and robustness. An awareness of the assumptions and approximations that a program makes is important, therefore, as the results may be misleading when the simplifications do not hold. The conclusions should always be checked against common sense.

A distinction is often made between 'systematic' and 'random' errors, closely followed by a discussion of 'accuracy versus precision'. From the viewpoint here, however, the issue is really one of considering whether all the significant sources of uncertainty have been taken into account in the analysis; they are dealt with in the same way, through the rules of probability theory. If there is a distinction to be made, it is that some uncertainties might usefully be reduced with appropriate calibration experiments.

The analysis of data is a dynamic process akin to a conversation. The underlying assumptions and approximations define a specific question that is being asked of the data. The immediate response tends to be of the type: "Well, that depends on what you mean by .....", and further work is required to formulate a well-defined question. When not satisfied with the results of an analysis, we need to reflect on the reasons for it: has an assumption, or approximation, been made that is not valid; or, as is more often the case, is there something else that we know, or believe to be true, that has not been put into the analysis? It's as if the question being answered is not quite the one we had in mind, but the response gives us guidance towards a better formulation.

## 16.8 Worked examples

> **(1)** Calculate the normalization constant, $A$, and mean of the exponential distribution $\mathrm{pdf}(x\,|\,\mu, I) = A\,e^{-x/\mu}$ for $x \geqslant 0$.

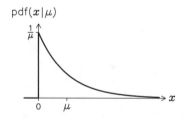

pdf$(x\,|\,\mu)$

$$\int_{-\infty}^{+\infty} \mathrm{pdf}(x\,|\,\mu, I)\,\mathrm{d}x = A\int_0^{\infty} e^{-x/\mu}\,\mathrm{d}x$$

$$= A\underbrace{\left[-\mu\,e^{-x/\mu}\right]_0^{\infty}}_{\mu} = 1 \qquad \therefore\ A = \frac{1}{\mu}$$

$$\mathbb{E}(x) = \int_{-\infty}^{+\infty} x\,\mathrm{pdf}(x\,|\,\mu, I)\,\mathrm{d}x$$

$$= \int_0^{\infty} x\,\frac{e^{-x/\mu}}{\mu}\,\mathrm{d}x$$

$$= \underbrace{\left[-x\,e^{-x/\mu}\right]_0^{\infty}}_{0} + \int_0^{\infty} e^{-x/\mu}\,\mathrm{d}x \qquad \therefore\ \mathbb{E}(x) = \mu$$

> **(2)** Show that both the mean and variance of the Poisson distribution, $\mathrm{prob}(r\,|\,\lambda, I)$ of eqn (16.21), is equal to $\lambda$.

$$\mathbb{E}(r) = \sum_{r=0}^{\infty} r \, \text{prob}(r|\lambda, I) = 0 + \sum_{r=1}^{\infty} r \, \text{prob}(r|\lambda, I)$$

$$= \sum_{r=1}^{\infty} \frac{\lambda^r e^{-\lambda}}{(r-1)!} \qquad\qquad \text{prob}(r|\lambda, I) = \frac{\lambda^r e^{-\lambda}}{r!}$$

$$= \lambda e^{-\lambda} \sum_{r=1}^{\infty} \frac{\lambda^{r-1}}{(r-1)!} = \lambda e^{-\lambda} \underbrace{\sum_{j=0}^{\infty} \frac{\lambda^j}{j!}}_{e^\lambda} \qquad \therefore \; \underline{\mathbb{E}(r) = \lambda}$$

$$\mathbb{E}(r^2) = \sum_{r=0}^{\infty} r^2 \, \text{prob}(r|\lambda, I)$$

$$= \sum_{r=1}^{\infty} \frac{r \, \lambda^r e^{-\lambda}}{(r-1)!}$$

$$= \lambda e^{-\lambda} \sum_{r=1}^{\infty} \frac{\big[(r-1)+1\big]\lambda^{r-1}}{(r-1)!}$$

$$= \lambda e^{-\lambda} \left[ \lambda \underbrace{\sum_{r=2}^{\infty} \frac{\lambda^{r-2}}{(r-2)!}}_{e^\lambda} + \underbrace{\sum_{r=1}^{\infty} \frac{\lambda^{r-1}}{(r-1)!}}_{e^\lambda} \right] = \lambda(\lambda+1)$$

$$\text{var}(r) = \mathbb{E}(r^2) - \big[\mathbb{E}(r)\big]^2 = \lambda(\lambda+1) - \lambda^2 \qquad \therefore \; \underline{\text{var}(r) = \lambda}$$

---

**(3)** Let $\{d_k\}$ be a set of N independent measurements ($k=1$ to N), with corresponding normal errors $\{\sigma_k\}$, of a quantity $\mu$. Find an expression for the most probable value of $\mu$, and a measure of its uncertainty, in terms of the $\{d_k\}$ and $\{\sigma_k\}$.

---

Following the analysis of a very similar problem in section (16.4), which was for the slightly simpler case of $\sigma_k = \sigma$, we obtain

$$L = \ln\big[\text{pdf}(\mu|\{d_k, \sigma_k\}, I)\big] = C - \frac{1}{2}\sum_{k=1}^{N} \frac{(d_k - \mu)^2}{\sigma_k^2}$$

instead of eqn (16.26). This is, in fact, just the least-squares expression of eqn (16.36) with $f_k = \mu$. For the most probable value of $\mu$ (given the data), $\mu_0$,

$$\frac{dL}{d\mu}\bigg|_{\mu_0} = \sum_{k=1}^{N} \frac{d_k - \mu_0}{\sigma_k^2} = 0 \qquad \Rightarrow \qquad \sum_{k=1}^{N} \frac{d_k}{\sigma_k^2} = \mu_0 \sum_{k=1}^{N} \frac{1}{\sigma_k^2}$$

This leads to the inference being summarized by $\mu = \mu_0 \pm \sigma_\mu$ where

$$\frac{1}{\sigma_\mu^2} = -\frac{d^2L}{d\mu^2}\bigg|_{\mu_0} = \sum_{k=1}^{N}\frac{1}{\sigma_k^2}$$

In other words, $\mu_0$ is given by a *weighted mean* of the data

$$\mu = \frac{1}{W}\sum_{k=1}^{N}w_k\,d_k \pm \frac{1}{\sqrt{W}}\ , \qquad \text{where}\ \ W = \sum_{k=1}^{N}w_k \qquad (16.41)$$

with the weight of the $k^{\text{th}}$ measurement being inversely proportional to its variance, $w_k = 1/\sigma_k^2$.

If the data were all subject to the same noise level, so that $\sigma_k = \sigma$, then eqn (16.41) should reduce to the earlier result of eqn (16.27). With $w_k = 1/\sigma^2$ and $W = N/\sigma^2$, this simplification is fulfilled as required.

---

**(4)** Let $\{d_k\}$ be a set of N independent measurements ($k = 1$ to N) with corresponding normal errors $\{\sigma_k\}$, and associated coordinates $\{x_k\}$, pertaining to a linear relationship: $d_k = f_k \pm \sigma_k$ where $f_k = m\,x_k + c$. Find expressions for the least-squares estimates of the gradient $m$ and intercept $c$, and their uncertainties.

---

With the simplifying approximations discussed in section (16.4), the least-squares assignment for the posterior probability of the gradient and intercept of the straight line model is

$$L = \ln\big[\mathrm{pdf}(m,c\,|\,\{d_k,\sigma_k\},I)\big] = C - \frac{1}{2}\sum_{k=1}^{N}\frac{(m\,x_k + c - d_k)^2}{\sigma_k^2}$$

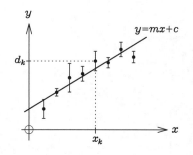

The optimal estimates, $m_0$ and $c_0$, are given by the solution to the pair of linear simultaneous equations resulting from the condition that

$$\frac{\partial L}{\partial m}\bigg|_{m_0,c_0} = 0 \qquad \text{and} \qquad \frac{\partial L}{\partial c}\bigg|_{m_0,c_0} = 0$$

The task is essentially the same as the minimization of the function $Q(m,c)$ in worked example (8) of chapter 10, where $L = C - \frac{1}{2}Q$, albeit with a change of notation ($X_i$ and $Y_i$ instead of $x_k$ and $d_k$) and a simplification ($\sigma_k = 1$). It yields

$$\begin{pmatrix} m_0 \\ c_0 \end{pmatrix} = \begin{pmatrix} \alpha & \beta \\ \beta & W \end{pmatrix}^{-1}\begin{pmatrix} \lambda \\ \mu \end{pmatrix}$$

where $\alpha$, $\beta$, $\lambda$, and $\mu$ are related to the data by

$$w_k = \frac{1}{\sigma_k^2} \qquad \alpha = \sum_{k=1}^{N}w_k\,x_k^2\,, \quad \beta = \sum_{k=1}^{N}w_k\,x_k\,, \quad \lambda = \sum_{k=1}^{N}w_k\,x_k\,d_k\,, \quad \mu = \sum_{k=1}^{N}w_k\,d_k$$

and W is defined in eqn (16.41).

The reliability of the estimate is specified by the covariance matrix, which is given by the inverse of the Hessian matrix $\mathbb{H}$

$$\mathbb{H} = -\begin{pmatrix} \partial^2 L/\partial m^2 & \partial^2 L/\partial m\,\partial c \\ \partial^2 L/\partial c\,\partial m & \partial^2 L/\partial c^2 \end{pmatrix} = \begin{pmatrix} \alpha & \beta \\ \beta & W \end{pmatrix}$$

That is to say

$$\alpha > 0, \quad W > 0, \quad \alpha W > \beta^2$$

$$\begin{pmatrix} \mathbb{E}\left[(\delta m)^2\right] & \mathbb{E}\left[\delta m\,\delta c\right] \\ \mathbb{E}\left[\delta c\,\delta m\right] & \mathbb{E}\left[(\delta c)^2\right] \end{pmatrix} = \begin{pmatrix} \alpha & \beta \\ \beta & W \end{pmatrix}^{-1} = \frac{1}{D}\begin{pmatrix} W & -\beta \\ -\beta & \alpha \end{pmatrix}$$

where $\delta m = m - m_0$, $\delta c = c - c_0$ and $D = \det(\mathbb{H}) = \alpha W - \beta^2$. Hence, the inference is summarized by

$$m_0 = \frac{W\lambda - \beta\mu}{D} \pm \sqrt{\frac{W}{D}} \quad \text{and} \quad c_0 = \frac{\alpha\mu - \beta\lambda}{D} \pm \sqrt{\frac{\alpha}{D}} \qquad (16.42)$$

and a correlation coefficient (between $m$ and $c$) of $-\beta/\sqrt{\alpha W}$.

---

**(5)** Given that $x = x_0 \pm \sigma_x$, where $0 < \sigma_x \ll x_0$, what can be said about (a) the reciprocal $1/x$, and (b) the logarithm $\ln x$?

---

(a)  Let $y = f(x) = \dfrac{1}{x} \approx y_0 \pm \sigma_y$

But $y = \underbrace{f(x_0)}_{y_0} + \underbrace{(x-x_0)}_{\delta x}\dfrac{df}{dx}\bigg|_{x_0} + \cdots \approx y_0 - \dfrac{\delta x}{x_0^2}$    $\delta y = y - y_0$

$\therefore y_0 = \dfrac{1}{x_0}$  and  $\underbrace{\mathbb{E}\left[(\delta y)^2\right]}_{\sigma_y^2} \approx \dfrac{\mathbb{E}\left[(\delta x)^2\right]}{x_0^4}$  $\Rightarrow$  $\dfrac{1}{x} \approx \dfrac{1}{x_0} \pm \dfrac{\sigma_x}{x_0^2}$

(b)  Let $z = g(x) = \ln x \approx z_0 \pm \sigma_z$

But $z = \underbrace{g(x_0)}_{z_0} + \underbrace{(x-x_0)}_{\delta x}\dfrac{dg}{dx}\bigg|_{x_0} + \cdots \approx z_0 + \dfrac{\delta x}{x_0}$    $\delta z = z - z_0$

$\therefore z_0 = \ln x_0$  and  $\sigma_z^2 \approx \dfrac{\sigma_x^2}{x_0^2}$  $\Rightarrow$  $\underline{\ln x \approx \ln x_0 \pm \dfrac{\sigma_x}{x_0}}$

A more rigorous error propagation procedure would entail a formal change of variables, from $\mathrm{pdf}(x|I)$ to $\mathrm{pdf}(y|I)$ where $y = f(x)$, with a Jacobian:

$$\mathrm{pdf}(x|I) = \mathrm{pdf}(y|I) \times \left|\dfrac{df}{dx}\right|$$

**(6)** A block of aluminium of mass $m = 453 \pm 1\,\text{g}$ absorbed $q = 5257 \pm 3\,\text{J}$ of heat and its temperature rose by $\theta = 12.6 \pm 0.1\,\text{K}$. Calculate the specific heat capacity $c \approx c_0 \pm \sigma_c$ of aluminium, where $q = mc\theta$.

Let $\quad c = f(q, m, \theta) = \dfrac{q}{m\theta} \approx c_0 \pm \sigma_c \quad$ and $\quad \mathbf{x}_0 = (q_0, m_0, \theta_0)$

$$\therefore\ c = \underbrace{f(\mathbf{x}_0)}_{c_0} + \underbrace{(q - q_0)}_{\delta q} \left.\frac{\partial f}{\partial q}\right|_{\mathbf{x}_0} + \underbrace{(m - m_0)}_{\delta m} \left.\frac{\partial f}{\partial m}\right|_{\mathbf{x}_0} + \underbrace{(\theta - \theta_0)}_{\delta\theta} \left.\frac{\partial f}{\partial \theta}\right|_{\mathbf{x}_0} + \cdots$$

$$\approx \underbrace{\frac{q_0}{m_0 \theta_0}}_{c_0} + \frac{\delta q}{m_0 \theta_0} - \frac{q_0\, \delta m}{m_0^2 \theta_0} - \frac{q_0\, \delta\theta}{m_0 \theta_0^2}$$

$$\therefore\ \underbrace{c - c_0}_{\delta c} \approx \frac{q_0}{m_0 \theta_0} \left[ \frac{\delta q}{q_0} - \frac{\delta m}{m_0} - \frac{\delta\theta}{\theta_0} \right]$$

$\mathbb{E}\left[\delta q\, \delta m\right] = 0 \quad$ etc. $\qquad \Rightarrow \quad \underbrace{\mathbb{E}\left[(\delta c)^2\right]}_{\sigma_c^2} \approx c_0^2 \left[ \frac{\mathbb{E}\left[(\delta q)^2\right]}{q_0^2} + \frac{\mathbb{E}\left[(\delta m)^2\right]}{m_0^2} + \frac{\mathbb{E}\left[(\delta\theta)^2\right]}{\theta_0^2} \right] + 0$

$$\text{i.e.}\quad \sigma_c \approx |c_0| \sqrt{\left(\frac{\sigma_q}{q_0}\right)^2 + \left(\frac{\sigma_m}{m_0}\right)^2 + \left(\frac{\sigma_\theta}{\theta_0}\right)^2} \quad \text{where}\quad c_0 = \frac{q_0}{m_0 \theta_0}$$

$$\therefore\ c_0 = \frac{5257}{453 \times 12.6} = 0.921\,\text{J}\,\text{g}^{-1}\,\text{K}^{-1}$$

$$\text{and}\quad \sigma_c \approx 0.921 \sqrt{\left(\frac{3}{5257}\right)^2 + \left(\frac{1}{453}\right)^2 + \left(\frac{0.1}{12.6}\right)^2} = 0.008\,\text{J}\,\text{g}^{-1}\,\text{K}^{-1}$$

i.e.    Specific heat capacity $\approx 921 \pm 8\,\text{J}\,\text{kg}^{-1}\,\text{K}^{-1}$

# Index